Our Movie Houses

Television and Popular Culture

Robert J. Thompson, *Series Editor*

Other titles in Television and Popular Culture

Athena's Daughters: Television's New Women Warriors
    Frances Early & Kathleen Kennedy, eds.

Framework: A History of Screenwriting in the American Film, 3rd ed.
    Tom Stempel

"Something on My Own": Gertrude Berg and American Broadcasting, 1929–1956
    Glenn D. Smith, Jr.

Starting Your Television Writing Career: The Warner Bros. Television Writers Workshop Guide
    Abby Finer & Deborah Pearlman

Watching TV
    Harry Castleman & Walter Podrazik

The West Wing: The American Presidency as Television Drama
    Peter C. Rollins & John E. O'Connor

Our Movie Houses

A History of Film & Cinematic
Innovation in Central New York

NORMAN O. KEIM

WITH DAVID MARC

Syracuse University Press

First Edition 2008
08   09   10   11   12   13   6   5   4   3   2

Unless otherwise stated, all illustrations courtesy of Norman O. Keim.

The paper used in this publication meets the minimum requirements of
American National Standard for Information Sciences—Permanence
of Paper for Printed Library Materials, ANSI Z39.48–1984.∞™

For a listing of books published and distributed by Syracuse University Press,
visit our Web site at SyracuseUniversityPress.syr.edu.

ISBN-13: 978-0-8156-0896-7   ISBN-10: 0-8156-0896-9

Library of Congress Cataloging-in-Publication Data

Keim, Norman O.
Our movie houses : a history of film and cinematic innovation in Central New York /
Norman O. Keim. — 1st ed.
p. cm. — (Television and popular culture)
ISBN 978-0-8156-0896-7 (hardcover : alk. paper)
1. Motion pictures—New York (State)—History.   2. Motion picture theaters—New York
(State)—History.   I. Title.
PN1993.5.U77K45 2008
384'.8'097476—dc22   2008005418

Manufactured in the United States of America

I wish to dedicate this book to my wife, Gloria,
who has endured years of trudging around central New York,
who has experienced countless nights of me working
in my "cave," and who has sent all of my e-mails.
Still she supports me in all of my endeavors.

Norman O. Keim

This book is published with the generous assistance of grants from

The Bleier Center for Television and Popular Culture

The Central New York Community Foundation

The Rosamond Gifford Charitable Corporation

NORMAN O. KEIM was United Ministries chaplain at Syracuse University's Hendricks Chapel from 1961 until 1973. He was adjunct professor of film studies at the university's Newhouse School of Journalism from 1973 until 1985, and founder/director of the SU Film Studies Center. He also created Film Forum, a three-night-per-week presentation of art films, both on and off campus, from 1967 until 1980. In addition, he served as an educational consultant providing career counseling at the Regional Learning Center in Syracuse, New York.

DAVID MARC is the author of five previous book and more than 250 articles for periodicals and reference works. He currently serves as associate editor of *Syracuse University Magazine* and is an editorial board member of *Television Quarterly*, published by the National Academy of Television Arts and Sciences (New York). Two of his books, *Prime Time, Prime Movers* (cowritten with Robert J. Thompson) and *Bonfire of the Humanities*, were published by Syracuse University Press.

❋ Contents ❋

Illustrations   *xi*

Color Plates   *xv*

Tables   *xvii*

Acknowledgments   *xix*

PART ONE **History**

1. Origins of American Film in Central New York State   3

2. The Nickelodeon Era, 1909–1919   22

3. The Roaring Twenties
   Movie Palaces and Theater Organs   38

4. The Great Depression and the War
   Talkies, Double Features, and Dish Night   57

5. End of the Studio Era
   TV, Widescreen Projection, and Drive-ins   65

6. The Schine and Kallet Circuits
   Exhibition in Central New York   72

7. Neighborhood Theaters in the City of Syracuse   87

PART TWO **People**

8. Cinema Figures with Links to Central and Upstate New York
   A Biographical Dictionary   95

Appendixes:

A. Theater Organs   141

B. Drive-in Theaters in Central New York,
   Past and Present   145

C. The Schine and Kallet Theater Chains
   in Central New York   147

D. Central New York Movie Theaters   157

❄ Illustrations ❄

1. Pamphlet on early film technology by William and Antonia Dickson   4
2. Eastman House kinetoscope model   5
3. Edison's "Black Maria" of 1893, America's first film studio   6
4. Edison kinetoscope model   7
5. The mutograph, the original biograph camera   8
6. The C. E. Lipe Machine and Tool Works in Syracuse, New York   9
7. Earliest known mutoscope, 1897   10
8. Later mutoscope encased in steel, c. 1900   11
9. Showbill for vitascope demonstration at Koster
   and Bial's Music Hall, 1896   12
10. Eugene Lauste's eidoloscope (or panoptikon) projector, 1896   13
11. K.M.C.D. Syndicate, 1895   14
12. Movable studio on the roof at 841 Broadway in Manhattan, 1897   15
13. Edison Film catalog cover, 1903   16
14. The Bastable block in Syracuse, which housed two theaters   17
15. Program for the Bastable Theatre in Syracuse, 1898   18
16. Grand Opera House in Syracuse   19
17. Advertising card for Syracuse's Grand Opera House, 1914   20
18. Alhambra Theatre, Syracuse   21
19. Star Theatre, Syracuse, early 1920s   23
20. Crescent Theatre, Syracuse   23
21. Program for Temple Theatre, Syracuse, 1928   24
22. Savoy Theatre, Syracuse   25
23. Novelty Theatre, Syracuse   26
24. Empire Theatre, Syracuse   27
25. Astor Theatre, formerly the Empire, Syracuse   28
26. Strand Theatre, Syracuse   28
27. Eckel Theatre, Syracuse   29
28. Arcadia Theatre, Syracuse   30
29. Avon Theatre, Syracuse   30

30. Lyceum Theatre, Syracuse   *31*

31. Happy Hour Theatre, Syracuse   *32*

32. Early film star Pearl White and friend   *33*

33. Wharton Studios in Ithaca   *33*

34. Streetcar plunging into an Ithaca gorge in
    a 1914 serial   *34*

35. Film stars Creighton Hale, Pearl White, and Lionel Barrymore
    at Wharton Studios in Ithaca   *35*

36. The Simplex projector, developed in 1914   *36*

37. RKO Keith's, Syracuse   *39*

38. B. F. Keith's program   *40*

39. B. F. Keith's vaudeville theater, Syracuse   *41*

40. Acme Theatre, Syracuse   *42*

41. Palace Theatre, Eastwood   *42*

42. Program for "Sonya's Search for the Christmas Star"   *43*

43. Brighton Theatre, Syracuse   *44*

44. Riviera Theatre, Syracuse   *44*

45. Crowd scene outside Riviera Theatre   *45*

46. Riviera Theatre interior   *45*

47. The Paramount, Syracuse   *46*

48. Loew's State Theatre, Syracuse   *47*

49. Loew's State Theatre interior   *47*

50. Syracuse's "Broadway Row"   *48*

51. System Theatre, Syracuse   *49*

52. Wurlitzer organ   *51*

53. The State Theatre in Ithaca   *52*

54. Marr and Colton Company advertisement   *53*

55. Theodore Case at Casowasco   *58*

56. Case and his assistant, Sponable, in lab   *59*

57. Interior of Case Research Lab   *60*

58. Fox-Case camera   *61*

59. Kallet Drive-in, Camillus   *70*

60. The Glove Theatre in Gloversville   *73*

61. Star Theatre, Oriskany Falls   *74*

62. John Eberson, ca. 1938   *75*

63. Schine Auburn Theatre, Auburn, ca. 1938   *77*

64. Foyer of the Schine Auburn Theatre   *77*

65. Chromed concession stand in the lobby of the Schine Auburn Theatre   *78*

66. Auditorium of the Schine Auburn Theatre   *79*

67. Strand Theatre, Rome   *80*

68. Star Theatre, Rome   *80*

69. Capitol Theatre, Rome   *82*

70. Farman's Theatre, Warsaw   *83*

71. O-At-Ka Theatre, Warsaw   *84*

72. Franklin Theatre, Syracuse   *84*

73. Regent Theatre, Syracuse   *85*

74. Shoppingtown Theatre, Dewitt   *86*

75. Genesee Theatre, Syracuse   *86*

76. Globe Theatre, Syracuse   *89*

77. Cameo Theatre, Syracuse   *90*

78. Community Theatre, Solvay   *91*

79. King Baggott, center, with Josephine Victor, left, and Maidel Turner, c. 1900   *98*

80. Louise Brooks   *101*

81. Pearl S. Buck, ca. 1932   *103*

82. Jackie Coogan   *107*

83. Norma Shearer Thalberg and her husband, Irving Thalberg, 1929   *126*

84. Stereograph of Mark Twain (Samuel Langhorne Clemens), ca. 1907   *129*

85. Clara Bow   *132*

86. Pearl White, ca. 1916   *134*

87. James Theatre, Utica   *152*

88. Uptown Theatre, Utica   *152*

89. Idle Hour Theatre, Rome   *162*

90. Franjo Theatre, Boonville   *162*

91. Quirk Theatre, Fulton   *164*

❧ Color Plates ❧

Following page 92

  1. Weiting Opera House

  2. Empire Theatre brochure

  3. Turn Hall

  4. Ladies' Room, B. F. Keith's Theatre

  5. Foyer, B. F. Keith's Theatre

  6. Boxes and balcony, B. F. Keith's Theatre

  7. Westcott Cinema today

  8. Brighton Theatre today

  9. Grand Staircase, Loew's Theatre

10. Auditorium from stage, Loew's Theatre

11. Grand Promenade today, Loew's Theatre

12. Fox Movietone truck

13. Strand Theatre, Brockport

14. Kallet Theatre, Oneida

15. *The Gorgeous Hussy*

16. *Rome Haul*

17. *Drums Along the Mohawk*

18. *Chad Hanna*

19. *Red River*

20. *Rio Bravo*

21. *Sergeant York*

22. *David Harum*

23. *Tess of the Storm Country*

24. *The Shepherd of the Hills*

❈ Tables ❈

Table A.1. Theater Organs in Central New York   *141*

Table A.2. Organ Manufacturing Companies   *143*

Table B.1. Operating Drive-in Movie Theaters
    in Central New York, 2007   *145*

Table B.2. Sampling of Closed Drive-in Movie Theaters
    in Central New York   *146*

Table C.1. Schine Theaters in New York State   *147*

Table C.2. The Kallet Theater Chain in New York State   *153*

Table D.1. Theaters in the City of Syracuse by Location   *157*

Table D.2. Other Central and Upstate New York Movie
    Theaters by Town   *163*

❉ Acknowledgments ❉

I have been collecting materials for more than twenty years and have not always kept the best records, so participants may have to excuse some omissions. But the following institutions and people come immediately to mind: Syracuse University Library (Barb Opar, Sue Miller, Dave Broda, Edward Galvin, Mary O'Brien), Onondaga Historical Association (Dennis Connors), Syracuse Public Library, Baldwinsville Public Library, Cayuga Museum of History and Art and Case Research Lab Museum (Carrie Barrett) in Auburn, Chittenango Landing Canal Boat Museum (Bud Fenner), Friends of History in Fulton (Julie Yankowsky), Geddes Historical Society (Carrie Gannett), Kallet Theaters (Robert Kallet, Joe Pfeiffer, Jr.), Liverpool Museum, Rome Historical Society (Rob Avery), Schine Historical Museum (Richard Samrov) in Gloversville, Limestone Ridge Historical Society in Oriskany Falls, History Center in Tompkins County (Donna Eschenbrenner), Utica Public Library, and Warsaw (N.Y.) Historical Society (Thomas L. Cardwell). Many thanks to all.

Thanks also to James Barrett, Bob Chamberlin, Jack Chesebro, Louis and Myrtle Chomyk, John Denitto, Joseph Detor, Lorenzo Fernandez, Will Headlee, David Jenks, Allen Kosoff, Joseph Lampe, Anthony Langan, Bill Lipe, Nicki McCarthy, Karen Colizzi Noonan, Art Pierce, George Read, Shirley Reidenbaugh, Barb Rivette, Ed Stephenson, Robert Thompson, Barb Thomson, and Conrad Zurich.

A special thanks to Dick Case, who gave me the impulse to continue with this project.

Many thanks to Ellen Goodman, my acquisitions editor at Syracuse University Press, who saw this project through from beginning to its fulfillment, to my copy editor, Linda Cuckovich, to my project editor, Marian Buda, and to my designer, Vicky Lane.

Many thanks to the Central New York Community Foundation, Inc., President's Discretionary Fund (Peggy Ogden), the Rosamond Gifford Charitable Corporation (Heidi Holtz), and the Bleier Center for Television and Popular Culture (Robert J. Thompson) for their assistance in the underwriting of this publication.

Norman O. Keim

PART ONE

History

I

Origins of American Film in Central New York State

Histories of American cinema often begin in New Jersey, about fifteen miles west of the Hudson in West Orange. That was the location of the Edison Company laboratory, where William K. L. Dickson invented the kinetoscope in 1891. But this account begins several years earlier, some three hundred miles northwest, in Rochester, New York, at the industrial laboratory of another great inventor of the period, George Eastman. Already famous for his pivotal contributions to the development of still photography, Eastman registered a patent for perforated celluloid film in 1889, a project he had taken on at Edison's suggestion. This new type of film, suitable for recording moving pictures, did not mean much to the public. No one had ever seen a movie, and many people were just discovering the thrill of taking their first snapshots with Eastman's push-button cameras. But news of the breakthrough had an immediate impact on Dickson and others competing in an international race to record motion by means of photography. Without compatible film, the kinetoscope was no better than a computer with no software—a box of impressive-looking hardware capable of nothing. With Eastman's new product, it would introduce America and the world to movies.

Dickson's boss, Thomas Alva Edison, a genius at applying the principles of mass production to new technologies, was less knowledgeable on the subject of mass entertainment. Although he had built America's first film studio, the "Black Maria," a small structure covered with tar paper, at his West Orange complex in 1893, Edison was slow to see prospects for the kinetoscope's commercial success. Blinded, perhaps, by the fact that an underling had done the work, Edison did not believe the big black box with the little peephole worth including among the inventions he showcased to the world at the Chicago Columbian Exposition of 1893. Nor was he willing to pay a $150 fee that would have secured him exclusive world manufacturing rights to the kinetoscope and the kinetograph (the camera that made images for it). This bit of penny-pinching by the Wizard of Menlo Park would cost his heirs dearly, but would speed up the development of the movies by putting it in the hands of others.

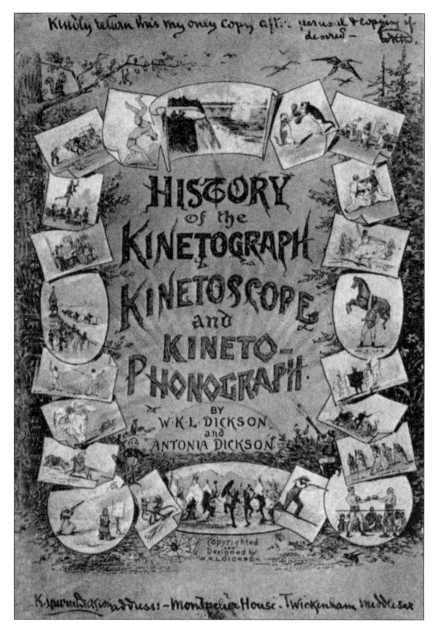

1. William Kennedy Laurie Dickson wrote this pamphlet with his sister, Antonia, in 1894.

George Eastman was born in Waterville, New York, about twenty miles west of Utica, and raised in Rochester. The death of his father ended his formal education at age fourteen, and the story of his rise to the ranks of the world's great industrial scientists rivals any in American history or fiction. But Eastman did not cultivate a "genius persona" in the style of Edison. He was by all accounts a

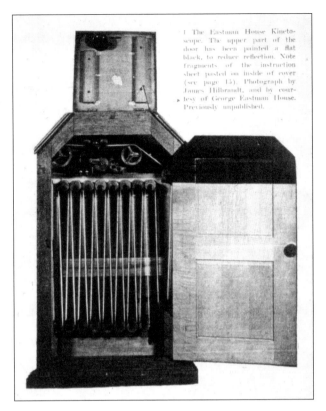

1 The Eastman House Kinetoscope. The upper part of the
door has been painted a flat
black, to reduce reflection. Note
fragments of the instruction
sheet pasted on inside of cover
(see page 15). Photograph by
James Hilbrandt, and by courtesy of George Eastman House.
Previously unpublished.

2. Eastman House kinetoscope. Courtesy of George
Eastman House.

shy man who measured his self worth by the positive impact of his work on the
lives of people. By producing easy-to-use handheld cameras and replacing metal
plates with film rolled on spools, Eastman expanded photography from an elite
specialist's technology to the most accessible and democratic of the arts to emerge
from the scientific advances of the century. His "Brownie" line of cameras dominated U.S. and international markets from its introduction in 1900 to the closing out of its final model in 1967, and his company, Eastman Kodak, helped the
movies become a leading form of mass entertainment through a continuing series
of technological advances, including the color film process that bears the Eastman name. At the Kodak Theater in Hollywood, home to the annual Academy
Awards ceremony since 2001, visitors to the George Eastman Room will find an
Oscar statuette on display, one of eight awarded to his Rochester company for its
contributions to American film.

Although unable to attend college, Eastman donated much of his personal
fortune to educational, cultural, and medical institutions. His residence, George
Eastman House, with holdings of more than twenty-five thousand films, as well as
collections of production stills, posters, and other paracinema, has made Rochester

3. The "Black Maria" was built in 1893 as a motion picture studio at Edison's laboratory in West Orange, New Jersey.

a world center of film research. The four institutions Eastman favored most in his giving reflect his central New York roots, although only one, the University of Rochester, whose medical school he founded, is located in the region. His philanthropy grew out of an old civic ethic that was indistinguishable to him from the self-evident duty of serving one's neighbors. The Massachusetts Institute of Technology received Eastman funding because of his belief that social progress is made possible by advances in technology, and although he had no formal connection to the school, he believed it a valuable engine for the improvement of society. Hampton Institute in Virginia (now Norfolk State University) and Tuskegee Institute in Alabama, two leading African American universities, received Eastman's support because they stirred his spirit. Born before the Civil War in the abolitionist heartland of central New York, Eastman formed his social philosophy in upstate towns where some families made their homes into stations on the Underground Railroad, and most families sent sons to the Union army. He saw Hampton and Tuskegee doing the unfinished work of the antislavery movement and responded.

History has not forgotten Eastman, but the primacy of his contributions to the making of cinema is often muted and sometimes forgotten, much as the two-word name of his company is often reduced to "Kodak." Today he is remembered more for his enduring philanthropic legacy than for his cinematic inventions and

4. Edison kinetoscope model. From the collections of The Henry Ford Museum.

improvements: the Eastman School of Music in Rochester, now among the world's finest; the dental clinics he founded in London, Paris, Rome, Brussels, and Stockholm, which helped introduce modern dental hygiene to millions; and his 1919 gift of ten million dollars in Kodak stock to company employees, which continues to pay their descendants dividends toward college educations and comfortable retirements. Like George Eastman, central New York has a long list of credits in the making of the contemporary world. So perhaps it is no wonder that the region's contributions to the making of motion pictures are as undervalued or forgotten as Eastman's own.

5. The original biograph camera, the mutograph, was used for the first time in early 1895 in Syracuse, New York. Courtesy of the Smithsonian Institution.

London, Paris, Berlin, New York City . . . and Syracuse?

While Edison dawdled in the making of movies in New Jersey, European competitors were hard at work. In London, Robert W. Paul was developing his theatrograph, whose machine movement was based on a medieval Swiss clock movement known as "the Maltese cross"; in France, the Lumière brothers, Auguste and Louis, were building their cinematograph, a sixteen-pound machine capable of photographing a motion picture and projecting it; and in Berlin, Max Skladanowsky was perfecting his bioscope. Edison had domestic competition as well, and to find it we must again look to central New York, where Herman Casler, a young machinist, was working on an early version of his biograph camera at the C. E. Lipe Machine and Tool Works on South Geddes Street in Syracuse.

The Lipe Works epitomized the transformation of backyard mechanical know-how into a shaping force of the contemporary world. At the turn of the twentieth century, Syracuse was a center of emerging technologies; automobiles, typewriters, electrical appliances, tool fabrication, and air conditioning were all under development by entrepreneurs in the city and surrounding region, and Lipe's shop gained a reputation as the city's "cradle of industry" as the work that went

6. The C. E. Lipe Machine and Tool Works, at 208 Geddes Street in Syracuse, New York, in the late nineteenth century. Courtesy of Barbara C. Lipe.

on there generated a steady succession of patents. A century ahead of its time, Lipe was a prototype for the start-up incubators and economic think tanks operated today by universities and regional industrial authorities. Enterprising young machinists came from everywhere to rent space and develop their ideas at Lipe, and they enjoyed the added benefit of close quarters with some of the best and the brightest emerging inventors, engineers, and industrial entrepreneurs of their generation. H. H. Franklin, for example, developed the first air-cooled automobile engine there. Lipe was so impressed that he became a collaborator, developing the transmission and differential for the Franklin automobile. Henry Ford traveled regularly to Syracuse from Dearborn on the New York Central Railroad's Niagara Rainbow, engaging the talent he encountered to manufacture parts for his first commercial production cars.

Henry Norton Marvin, a machine tool fabricator who had invented the first electric drill capable of penetrating rock, became acquainted with Herman Casler at Lipe, and the two became friends, collaborators, and business partners. Their invention, the mutoscope, was similar to the Edison Company's kinetoscope, offering a viewer a moving image cinema through a peephole. But the mutoscope created the image by a radically different process. A series of still photographs were placed on a wheel inside a box, which was operated by the viewer with an external hand crank. In essence, it was an automated version of an old parlor-room favorite,

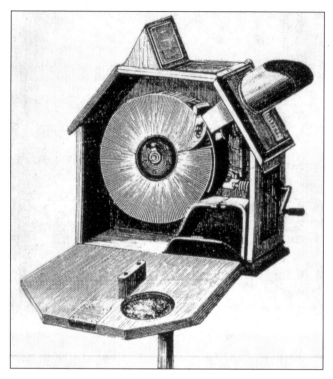

7. Earliest known version of mutoscope, from the *Scientific American*, April 17, 1897.

the flipbook. When viewed in quick succession, the photos stimulate the human capacity for persistence of vision, creating an illusion of motion.

The American Mutoscope and Biograph Company was founded by the inventors and their business partners to capitalize the project, and they formed a separate company, Marvin and Casler, as a licensed manufacturer. The mutoscope became the principal competitor to the kinetoscope. From 1895 to 1909, penny arcades around the country typically featured both machines among their attractions. Some viewers preferred the greater control over the speed of the image offered by the mutoscope's hand crank. Others preferred the steady pace offered by the kinetoscope, whose electric motor pulled the celluloid filmstrip over the light source. Eventually, the mutoscope proved more popular than its rival because it was able to introduce larger images without the distortion problems that plagued the kinetoscope.

During the arcade era, one machine served one customer at a time. The next step in development of the movies was to find a way to project a single large image onto a screen in a space where hundreds of customers could be served at once. Here again, Edison dawdled and penny-pinched. The kinetoscope was earning its keep—something he once thought unlikely—and he saw no reason to shake up

8. A later version of the mutoscope was encased in steel so that it could not be carried away, c. 1900.

the arcade business. Besides, efforts in West Orange to develop a film projector had not borne fruit. Edison was moved to act only after a rumor circulated that the Lumière brothers were coming to New York to demonstrate their motion picture projection system. He bought the rights to the phantoscope, a projector developed by Francis Jenkins and Thomas Armat in Washington, D.C. To cover up its origin, he introduced it to the public as the vitascope.

On April 23, 1896, Edison set up a screening at Koster and Bial's Music Hall, a vaudeville house on Herald Square in the heart of Manhattan. A twelve-minute presentation was made, consisting of ten kinetoscope film loops spliced together. The successful demonstration of the large image on the theater stage caused a sensation in the New York City press that was heard around the world. The Koster and Bial demonstration is cited in standard screen histories as the first projection of a movie to a public audience.

There is, however, evidence of an earlier screening—upstate. On March 28, 1896, some twenty-six days before the Edison event, Eugene Lauste offered a public demonstration of film projection at the Everson Building, 110 South Salina Street,

9. Show bill for the demonstration of Thomas Edison's vitascope at Koster and Bial's Music Hall, April 20, 1896.

in the heart of downtown Syracuse. Lauste developed his projector, the panoptikon (also known as the eidoloscope), as an employee of the Lamda Company, founded by Woodville Latham, a former Confederate army general, and operated by his sons, Grey and Otway. The Latham brothers, based in New York City, had worked for Edison selling kinetoscopes to arcades in various cities, including Syracuse. They had more than a passing familiarity with Edison technology, and, given Edison's penchant for bringing lawsuits in such situations, the Lathams may have planned the demonstration for Syracuse to escape his notice.

There is at least one striking similarity between the Syracuse and Manhattan events: the splicing of ten kinetoscope loops to make a twelve-minute movie. Lauste did it first—but did the Lathams steal the idea from Edison's plans or did Edison learn of the Syracuse screening and copy Lauste? Either is possible. As

10. Eugene Lauste's eidoloscope (or panoptikon), April 28, 1896. The eidoloscope projector was used for the first film showing in Syracuse, New York, in March 1896.

entertainment events, the Syracuse screening was far more ambitious. Lauste offered cinema to the public not as a novelty or a trick tacked on to a vaudeville show, but as a new type of commercial theatrical amusement, running continuous showings of the film from 10 A.M. to 10 P.M. for three full weeks.

The moving picture show at the Everson created a sensation—a sensation, that is, in Syracuse and surrounding counties. Learning of it, Edison announced a boycott of the Latham brothers for their unsanctioned use of the kinetoscope loops, as his lawyers planned a suit. The Latham brothers, lacking sufficient finances to develop Lauste's discovery—or to defend it from Edison's lawyers in court—soon passed from the scene. Freeman Galpin documented Lauste's 1896 Syracuse screening—arguably the very first theatrical movie engagement—in "A History of the Motion Pictures in Syracuse," published in *New York History* (September 1955).

11. The first meeting of all four members of the K.M.C.D. Syndicate—
H. N. Marvin, William Kennedy Laurie Dickson, Herman Casler, and
E. B. Koopman—took place on Marvin's lawn in Canastota, New York,
September 22, 1895.

The Biograph Company of Canastota, New York

Angered that Edison had not given him adequate credit for his work on the kinetoscope, W. K. L. Dickson left West Orange in 1895 to join Casler and Marvin. The pair had outgrown the Lipe factory and opened up a shop of their own, some ten miles east of Syracuse along the Erie Canal in Canastota, New York, where Henry Marvin's mother lived. Elias B. Koopman, a financier active in the entertainment business, joined the partnership, bringing with him George William "Billy" Bitzer, who had impressed him with ideas about what kind of content might work in the new medium. Bitzer would later become famous as D. W. Griffith's cinematographer, shooting more than nine hundred films in a Hollywood career that lasted until the 1940s. The quintet of Casler, Marvin, Dickson, Koopman, and Bitzer came together as founding partners of the American Mutoscope and Biograph Company, or Biograph, as it would eventually be known around the world. Responsibilities were parceled out: Casler would stay in Canastota to develop new technology. Marvin headed for Manhattan to build a studio on the roof of the Koopman building at 841 Broadway. Koopman would bring in investors and mind the books. It was up to Dickson and Bitzer to make the movies.

12. Photo of the movable studio built in late March 1897 on the roof at 841 Broadway in Manhattan, from the *Scientific American*, April 17, 1897.

Among the first hires made at Biograph was *Syracuse Journal* reporter Frank J. Marion, who probably was the first screenwriter ever hired as such. Marion teamed with Wallace McCutcheon, and the pair wrote many of Biograph's most memorable comedies, including *Troubles of a Manager of a Burlesque Show* (1904) and *Tom, Tom, the Piper's Son* (1905). The latter was rephotographed and examined in exquisite detail in *Tom, Tom, the Piper's Son* (1969) by Ken Jacobs, a Binghamton University professor. Marion left Biograph in 1905 to go west and become a founding partner of Kalem Pictures, one of the first Hollywood production houses. Marion made several important advances at Kalem, including production of one of the first films of the life of Jesus, *From the Manger to the Cross* (1912), which was also among the first films shot on location in the Holy Land.

Edison brought suit against Biograph, as he did against all competitors. But with Koopman's financial backing, the Biograph partners had adequate legal assistance in demonstrating the important differences between the mutoscope and the kinetoscope. Biograph won its case and emerged as Edison's only significant competitor in the penny arcade era, positioning the company for the age of projection.

Lauste's 1896 screening may or may not have influenced the emergence of Syracuse as an early center of film exhibition. Cold, snowy winters and a fascination for new gadgetry in a population thriving on high-tech manufacturing could not have hurt the rise of the movies. The first wave of commercial screenings in Syracuse, as elsewhere, were booked on the Edison model: as the final act in

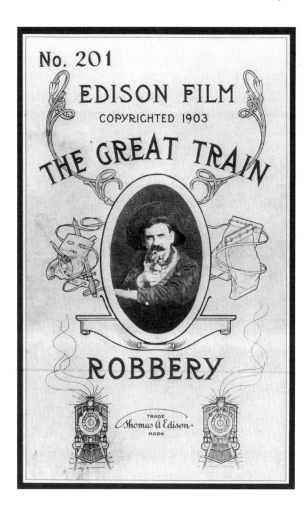

13. Cover of Edison Film Catalog No. 201, featuring *The Great Train Robbery,* 1903.

a show containing singers, dancers, comedians, magicians, trained animals, and contortionists. Lasting no longer than about twelve minutes, movies were known as chasers. Early fans thought of them as a refreshing beer following a stiff drink; the unimpressed saw them as an efficient way of getting people out of their seats to hasten preparation for the next show.

These first films from Edison and Biograph were merely minute-long scenes. The phenomenon of cinema—enjoying the power of the illusion—was entertainment enough. Narrative development was neither necessary nor, at the beginning, technically feasible. A movie was a horse running; a man sneezing; a baby in a bath; a scene at the shore; the New York Central's Empire State Express rounding a curve in a plume of smoke; a beautiful girl dancing. The Spanish-American War was captured on film. Audiences did not know that many of the Cuban war scenes were shot on the plains of New Jersey, nor would they have likely cared. Boxing matches and passion plays were early favorites at the box office—as was scandal. *The Kiss* (1896), a filmed scene from the Broadway play

14. Before burning down in a spectacular blaze in 1923, the Bastable block in Syracuse had two theaters: the Bastable (1893–1923) and the Standard (1913–23). The State Tower Building now stands on that site. Collection of the Onondaga Historical Association, Syracuse, New York.

The Widow Jones, raised eyebrows and helped alert the guardians of morality to the dangers of the movies. Comparing the stage and screen versions in the avant-guarde arts journal *The Chap-book*, the painter John Sloan wrote that even in the play "the spectacle of their prolonged pasturing on each other's lips was hard to bear. When only life-size it was pronounced beastly. But that was nothing to the present sight. Magnified to Gargantuan proportions and repeated three times over it is absolutely disgusting."

In Paris, Georges Méliès, a magician at heart as well by profession, dazzled audiences by filling the screen with dissolves, double exposures, and stop motion scenes. His fifteen-minute masterwork, *A Trip to the Moon* (1902), was one of the most loved and admired of early films. Méliès also pioneered color on screen by hand tinting and painting images right on the celluloid. In Dover, New Jersey, Edwin Porter, working for Edison, made *The Great Train Robbery* (1903), arguably the first action film *and* the first Western, all in one twelve-minute shot, available in black-and-white and hand-tinted versions. But France and New Jersey cannot claim all of the action during the acceleration of filmmaking that took place from 1900 to 1905. *Battle of the Yalu* (1904), purportedly a newsreel brought back from the front of the Russo-Japanese War, was in fact filmed by Billy Bitzer for

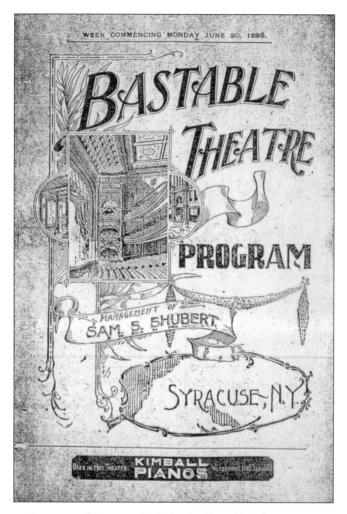

15. Program of June 13, 1898, for the Bastable Theatre in Syracuse, then under the management of Sam Shubert.

Biograph on the fields of St. John's Military Academy in Manlius, New York, a suburb of Syracuse. Colonel Verbeck, head of the academy, played a Russian general on horseback, and the military cadets—wearing authentic costumes—played the Russian and Japanese soldiers. The academy's two Japanese students got the lion's share of close-ups.

Early Movie Venues in Syracuse

The Wieting Opera House, on Clinton Square opposite the Erie Canal (see plate 1), was one of the premiere theaters in the East, and crucial to the city's reputation as a prime try-out spot for Broadway-bound plays. Lillian Russell, Victor Herbert, Helena Modjeska (the great Polish actress), and all the big stars of

16. The Grand Opera House on Syracuse's East Genesee Street stood across from City Hall. The block no longer exists.

the Broadway stage performed at the Wieting during the later decades of the nineteenth century, and the shows included many premieres. At the turn of the twentieth century, the Wieting began to book vaudeville shows—and booked the movies that came with them as chasers. Victor Herbert was said to love Syracuse for its pasta and beer. The third Wieting Opera House (after the original opera house and a subsequent one were destroyed by fire) was built on the site in 1897 and stood until 1932. As live theater went into decline, the Wieting held out the longest by switching to movies.

The Bastable Theatre, only a bit below the Wieting in standing, stood on the site of the current State Tower Building at Warren and Water Streets in Syracuse. In October 1897, a young theater manager named Sam Schubert booked a

17. Advertising card for the Grand Opera House for week of September 7, 1914. The Grand stood at 312 East Genesee Street in Syracuse. Courtesy of Richard G. Case.

film, projected on the veriscope system, of a heavyweight boxing match between Gentleman Jim Corbett and a fighter named Fitzsimmons. It filled every seat in the Bastable, and after that, all the legitimate houses in Syracuse were in line to show films. The Bastable burned in a spectacular blaze in 1923.

The Grand Opera House, 312 E. Genesee Street, was across from City Hall and the Yates Hotel, on a diagonal block of Genesee Street that no longer exists. It was designed by Archimedes Russell, who gave the theater's interior the identical ornate woodwork he had used in Syracuse University's Crouse College building. The Grand was the first theater in Syracuse lit completely by electricity. It was operated at the turn of the century by the Schubert family, and later became part of the B. F. Keith vaudeville circuit.

The Alhambra Theatre, 275 James Street, was a cavernous structure suited for conventions and mass meetings, but it did occasionally show films. It often featured musical concerts, with a two-reel comedy thrown in to fill the seats. Later it became a roller skating rink. A raging fire destroyed this historical building in 1954.

Dunfee's Comedy Theatre, 119 S. Clinton Street, lasted for only a few years at the turn of the century. It brought down the wrath of the city administration when

18. Although the Alhambra Theatre, at 275 James Street in Syracuse, was a cavernous building more suited for conventions and large meetings, it did show movies. It later became a roller skating rink before burning down, with the loss of some lives. Collection of the Onondaga Historical Association, Syracuse, New York.

it inaugurated Sunday showings. It reopened as the Lyceum for several years before it too burned down.

Film Under the Stars

Two outdoor theaters in the Syracuse area were built at the end of trolley lines by traction companies to stimulate business on summer weekends, and the price of admission was tied to a trolley ticket. They were rough-hewn structures, covered by canvas. Both were gone before 1920. Outdoor film showings were revived in downtown Syracuse as a feature of the Syracuse International Film Festival in the twenty-first century. The Lakeside Rustic at Maple Bay, on the west shore of Onondaga Lake, was known mainly for live musical theater productions. Films were shown as added attractions.

The Valley Theatre on Seneca Turnpike in Onondaga Valley was known primarily for live operas, but in 1899 it showed one of the first films of a passion play enactment, *The Horitz Passion Play* (1897), shot by an American crew in the tiny Czech village of Horitz, where it had been performed for centuries.

2

❧ The Nickelodeon Era, 1909–1919 ❧

Storefront movie theaters began replacing the early peephole devices in 1905, first in New York City and then across the country. The New York press dubbed these theaters "nickelodeons," combining the standard five-cent price of admission with "odeon," an ancient Greek theater with a roof (drama was normally performed outdoors in ancient civilizations). In 1907, Allen May opened Syracuse's first nickelodeon as an added attraction to his penny arcade at 334 South Salina Street. The public had been introduced to moving pictures at arcades, making them logical venues for presenting movies in their new and improved form—on a screen. May's venture did not last, but nickelodeons did. By 1909, ten were doing business in Syracuse. The Star Theatre, seating 312 in a storefront at 547 North Salina Street, became a downtown fixture, outlasting all its first-generation exhibition rivals before closing in 1939. The Star's owner, Gene Logan, was a carnival strongman and vaudevillian. His act included a stunt in which he lay on his back and an automobile was driven across his chest.

Logan was also a photography enthusiast, and he occasionally used his skill with a camera and his show business connections to get involved on the production side of the movie business. In 1923, during the shooting of *A Clouded Name* (a.k.a. *A Clouded Mind*) at the Calthrop mansion on the 3500 block of South Salina Street in Syracuse, Logan stood in for a missing cinematographer (credits list him as "Jean Logan"). Norma Shearer's performance in the film helped her win a long-term contract with MGM. In 1929, Logan picked up his camera professionally once again and traveled some twenty miles west of Syracuse to Auburn to capture rioting under way at the New York State Penitentiary for Fox Movietone News, Twentieth Century Fox's newsreel division. But the studio had other ideas in mind. A year later, Logan's footage of the Auburn riots came to the screen, uncredited, in *The Big House* (1930), a Fox prison movie starring Wallace Beery, which won two Oscars and was nominated for best picture.

In 1909, the Cahill brothers built the Crescent Theatre at 451 South Salina Street. The Crescent may have been the first theater in the country created exclusively for the exhibition of motion pictures. That claim is often made in standard film histories for the Regent Theatre at 116th Street and 7th Avenue in Harlem, which was built by Henry Marvin, one of the Biograph partners, some

19. The Star Theatre was a nickelodeon at 547 North Salina Street in Syracuse. Photo by Raymond La Rose for the *Syracuse Post Standard*, early 1920s. Courtesy of Richard G. Case.

20. The Crescent Theatre was the first large theater built on Syracuse's South Salina Street by the Cahill brothers. It stood next to Dey Brothers department store, about where the Galleries now stand. It lasted only through the silent movie era. Collection of the Onondaga Historical Association, Syracuse, New York.

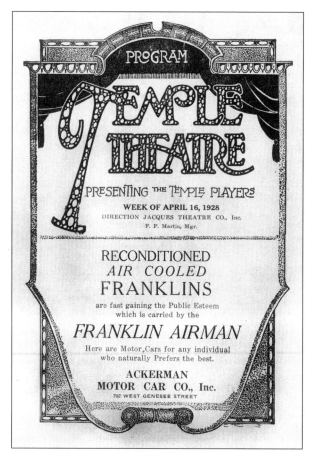

21. Temple Theatre program for the week of April 16, 1928. The Temple Theatre was the predecessor of the Paramount Theatre in Syracuse. Courtesy of Richard G. Case.

four years later. By the time the Marvin's downstate Regent opened, the Cahills were already at work on a second made-for-movies theater, the Temple, which opened in 1914 at 424 South Salina, across the street from the Crescent. The Paramount Theatre would rise on that site in 1929. A third Cahill cinema, the Savoy Theatre, went up on Warren Street in 1914, next to the Journal Newspaper Building, down the block from the Onondaga Hotel. The Savoy eventually became a burlesque house.

The Novelty Theatre, with 550 seats, stood at 213 West Fayette Street from 1909 to 1953. It showed popular serials, which were a big attraction for children, but the Novelty's reputation as a hangout for unsavory characters led many Syracuse mothers to make it off-limits. The joke around town about the Novelty was, "If you caught a rat you get your money back."

22. The Savoy Theatre was at 316 South
Warren Street next door to the *Syracuse Journal*
newspaper building and down the street from
the Onondaga Hotel. It later became a burlesque
house. Collection of the Onondaga Historical
Association, Syracuse, New York.

In 1911, the Gurney Real Estate Company built the Empire Theatre (1911–
55) at 468 South Salina Street (see plate 2). With fifteen hundred seats, it was
intended as a rival to the Wieting Opera House for live performances. The original Empire had a spacious lobby done in green and white Passanova marble, with
a dome of multicolored Tiffany glass. The predominant colors in the auditorium,
green and gold, were extensions of the lobby design: dark green carpeting in the
aisles and a lighter green on the side walls. But the Empire was built too late to
fulfill the vision of its designers. The glory days of the "legitimate stage" were
already numbered, as movie technology advanced to the point at which a photoplay could be equivalent in length to an evening at the theater. In 1915, the
Empire presented D. W. Griffith's *The Birth of a Nation* (1915), a prototype of the
modern feature film in length and form, and charged the unheard price of two
dollars for admission. The controversy surrounding the film's glorification of the
Ku Klux Klan, even in a town that had given so many of its sons to the Union

23. The Novelty Theatre, at 213 West Fayette Street in Syracuse, stood across the street from the Rivoli Theatre, where Pascale's Restaurant now stands. It has the typical nickelodeon front and must have been a real hole-in-the-wall, for the saying used to go that "If you caught a rat, you got your money back." Collection of the Onondaga Historical Association, Syracuse, New York.

army, only served to swell ticket sales. The Empire was renamed the Astor during its last years; the building was demolished in 1955.

The Syracuse Arena (1911–39), with an enormous capacity of six thousand seats, opened its doors in 1911 at 1140 South Salina Street, at the corner of Raynor Avenue, just west and downhill from the Syracuse University campus. Financed by R. M. Hazard, former owner of the Solvay Process Company in Solvay, New York, it was built as an indoor sports, entertainment, and convention hotspot, and at one time could boast the largest artificial ice skating rink in New York State. Movie exhibition was just one of the many types of events that took place at the Arena. In 1912, the Bull Moose Party, a breakaway splinter party of Republicans and Progressives, held its state convention there, nominating favorite son Theodore Roosevelt, who had served as president earlier in his career and was attempting a comeback. The party disappeared after Woodrow Wilson defeated TR in a three-way race. Boxing was popular at the Arena, and Jim Corbett, Jack Dempsey, and Sugar Ray Robinson were among the champions who

24. The Empire, at 468 South Salina Street, Syracuse, was a
beautiful theater built to rival the Wieting Opera House. It
showed *Birth of a Nation* in 1916. Later it was renamed the
Dewitt and, in its last years, the Astor. The building still stands,
though not as a theater. Collection of the Onondaga Historical
Association, Syracuse, New York.

fought there. In the 1920s, evangelist Billy Sunday brought his crusade there.
Located a short distance from where the Carrier Dome stands today, the Arena
was perhaps the closest thing central New York had during the early twentieth
century to an indoor venue of the Dome's size. It was a cavernous structure with
poor acoustics, but some films were shown there.

Downtown movie theater-building reached a peak of activity in 1914, when
many new theaters opened their doors. The Strand, built that year at the corner
of South Salina and Harrison Streets, opposite the Hotel Syracuse, was a beautiful
theater seating 1,600 people and presaging the great movie palaces of the 1920s.
It eventually became part of the Loew's chain, controlled by MGM, before closing

25. Syracuse's Empire Theatre was renamed the Astor in the early 1950s.
Courtesy of Barbara Thomson.

26. The Strand stood on the corner of Harrison and South Salina streets, across from
the Hotel Syracuse. It had an unusually large stage. Courtesy of Barbara Thomson.

27. The Eckel Theatre, at 216 East Fayette Street, was the earliest and the last of Syracuse's great downtown theaters. The Schines introduced Cinerama to central New York in this theater. Courtesy of Barbara Thomson.

in 1959. The Eckel Theatre, at 214–16 East Fayette Street at Montgomery Street, was built by descendants of Philip Eckel, one of the leaders of Syracuse's north-side German immigrant community. It was solely intended as a cinema, lacking any kind of stage area, and, like the Strand, seated 1,600. In the 1950s, the Eckel had a giant Cinerama screen, the only one of its kind in the city for many years. It closed in 1978. The Regent Theatre, at 820 East Genesee Street, was built in 1914 by Arthur Breese Merriman, the scion of two old-line Syracuse families. He also owned two other theaters built the same year—the Arcadia (with 550 seats) at South Salina and Colvin Streets, and the Franklin (with 400 seats) at 237 South Avenue—as well as the Alcazar on South Townsend Street (renamed Oakwood Avenue). Merriman was building the Avon Theatre on Hawley Avenue some years later when he suddenly went into Episcopal holy orders. He later served as rector of St. James Church in Skaneateles, New York, and of the Church of the Savior on James Street in Syracuse. The Regent, which had 1,000 seats, eventually became a "second-run exclusive" theater. This meant it would get films nine to twelve days after they opened downtown and two weeks before the other neighborhood theaters. The Regent was sold to Syracuse University in 1958, and the site became home to Syracuse Stage, one of upstate New York's leading theater companies. The Arcadia closed in 1931; the Franklin, in 1976.

28. The Arcadia Theatre was located at the intersection of South Salina Street and Colvin Street in a building that was later occupied by the Simmons Mortuary School. Collection of the Onondaga Historical Association, Syracuse, New York.

29. Syracuse's Avon Theatre, on Hawley Avenue.

30. The Lyceum Theatre stood at 245 West Fayette Street, almost at the corner of South Franklin Street and just around the corner from the New York Central Railroad station. Originally known as the Bijou from 1909 to 1912, it was renamed the Lyceum from 1912 to 1915. Collection of the Onondaga Historical Association, Syracuse, New York.

More than fifty theaters were built or converted to movie use in Syracuse during a three-year period: ten in 1912; twenty-three in 1913; and nineteen in 1914. Most were rebuilt storefronts, and many lasted only a year or two. Since not all advertised in the newspapers, it is difficult to comprehensively catalogue them. Some early theater operators simply put signs out front and opened their doors to an inquisitive public. As the movies grew in popularity, they were no longer strictly a downtown phenomenon. Neighborhood theaters were built on the main thoroughfares radiating from the center of the city, especially those with streetcar lines, and they became congregating points in their communities. Admission prices were lower in the neighborhoods, and for many moviegoers, downtown was reserved for special occasions.

The Arcadia was one of several theaters in Syracuse where patrons entered at the screen end and walked toward the back of the theater to take their seats. The Lyric at 258 Wolf Street was another of that design; the Simmons School of Embalming would later occupy the site. Turn Hall, on Syracuse's north side at 621 North Salina Street (see plate 3), was probably the only ethnic theater in Syracuse. During the silent era, it showed German and Italian films with original language title cards; during the sound era, it used the original soundtracks. The

31. The Happy Hour Theatre was at 224 North Salina
Street, across the street from where the Syracuse
newspaper building now stands. It later became the
Swan and still later the Midtown. Collection of the
Onondaga Historical Association, Syracuse, New York.

theater was in the headquarters of the German American Turn Verein, an immigrant social society. It burned down in a spectacular blaze in 1954.

What brought about the sudden surge of theaters during the mid-teen years? For one thing, films were getting better. D. W. Griffith was not only making sophisticated chase pictures for Biograph, but he was also developing the first dramatic film stars, including Mary Pickford, Lillian and Dorothy Gish, and Mae Marsh. Mack Sennett was turning out hilarious comedies with his troupe of stars, including the Keystone Kops, Mabel Normand, and, for a time, Charlie Chaplin, an English vaudevillian who learned the new medium from Sennett. Chaplin was readying himself to become the greatest movie star of his era.

Serials with female heroines attracted large female audiences to the movies. Pearl White, the star of the *Perils of Pauline*, was the first such heroine, and others followed, including Kathleen Williams, Helen Holmes, Ruth Roland, and Grace Cunard. *Exploits of Elaine*, starring Pearl White, was made at the Wharton Studios

32. Early film star Pearl White and friend pose by Cayuga Lake. Courtesy of the History Center in Tompkins County, New York.

33. The tower at Renwick (now Stewart) Park (since torn down) housed the lakeside offices of Wharton Studios in Ithaca between 1910 and 1920. Courtesy of the History Center in Tompkins County, New York.

34. A retired streetcar plunges to the bottom of a gorge near Ithaca, New York, for a 1914 serial. The president of the Ithaca Transit Company decorated his office with a large picture of the crash. Courtesy of the History Center in Tompkins County, New York.

in Stewart Park in Ithaca, New York, at the southern tip of Cayuga Lake. Wharton directors loved the Ithaca gorges, and they particularly enjoyed sending automobiles and streetcar trolleys catapulting over the rims. More than seventy films were made by the Wharton Brothers and other companies in Ithaca between 1914 and 1920. Irene Castle, Creighton Hale, Pearl White, Lionel Barrymore, King Baggot, Warner Oland, Oliver Hardy, Francis X. Bushman, Beverly Bayne, Milton Sills, and Norma Talmadge were among the many stars who journeyed from New York City to Ithaca on the Black Diamond, the Lehigh Valley Railroad's crack first-class New York–Buffalo train that traveled the Finger Lakes route.

Several Ithacans were able to capitalize on the town's exciting new industry. Louis Wolheim, a Cornell mathematics graduate who was hanging on in Collegetown as a part-time math instructor, found a career as a film actor. Marked for bad-guy character parts by a broken nose he received as a halfback on the Big Red football team, Wolheim appeared in more than fifty films between 1914 and 1931, including *All Quiet on the Western Front* (1930). Cinematographer Ray June, an Ithaca native who did not attend college, shot his first film, *The New Adventures of J. Rufus Wallingford* (1915), in his hometown. During a career that took him to Hollywood and lasted more than forty years, June was cinematographer for 166 commercial releases. He had just finished shooting *Houseboat* (1958), starring Cary Grant and Sophia Loren, when he died suddenly. Harold MacGrath, a screenwriter and novelist from Syracuse, wrote scripts for several serials shot in Ithaca and caught the attention of William Randolph Hearst; Hearst then

35. At the Wharton Studios at Ithaca's Renwick Park, film stars
Creighton Hale, Pearl White, and Lionel Barrymore sit beneath
a sign announcing, "Only Those Engaged in the Cast Permitted
in this Studio." Large crowds forced the Whartons to close the
studio grounds to the public. Courtesy of the History Center in
Tompkins County, New York.

financed MacGrath's first feature script, *The Vengeance that Failed* (1912), also
made in Ithaca. Twenty-eight MacGrath scripts reached the screen before the
writer's death in 1932.

The decade beginning in 1910 saw film companies at work across New York
State. The Kalem Company, founded in 1907 by Samuel Long, George Kleine,
and Syracuse newspaperman Frank Marion, was located on West 24th Street in
Manhattan. One of its first productions was a fifteen-minute version of *Ben Hur*,
made without the permission of the novel's author or publisher. Kalem was sued by
Harper's Publishing and lost in a landmark decision.

Some of the oldest silent film production facilities are still in use. The Vitagraph Studios in Brooklyn, at Avenue M and East 16th Street, began making silent

SIMPLEX
NOW READY
FOR DELIVERY

FULL PARTICULARS WILL BE FURNISHED
UPON REQUEST TO ANY BRANCH
OF THE NATIONAL THEATRE SUPPLY CO.
OR TO THE
INTERNATIONAL PROJECTOR CORPORATION
90 GOLD STREET, NEW YORK

36. In 1914 the Simplex projector eliminated stutter from film screenings.

pictures in 1907. It is perhaps the oldest continuously operating studio in the world, spanning three media during more than a century of service: Vitagraph silent films, 1907 and 1925; Warner Brothers talkies, 1925–53; and NBC television programs, 1953–present. Biograph's East Harlem facility, at 106th Street and Park Avenue, has gone through many regime changes, but is flourishing today as Metropolis Studios, an all-digital television production house. The Famous Players–Lasky Studio in Astoria, Queens, was busy throughout the silent era, and several important early sound films, including the Marx Brothers's *The Cocoanuts* (1928), were made there. After being closed for a period, it reopened as Kaufman Astoria Studios in 1977 and is among the busiest production centers in the country. Woody Allen has made many of his films there. Television series include *Sesame Street*.

While some of the early studios used the wilds of New Jersey for outdoor shooting, Biograph directors favored an unspoiled area of Orange County, near Cuddebackville, New York, north of Port Jervis. In all, some twenty Biograph films were shot there between 1909 and 1912. D. W. Griffith, who was particularly fond of the old Delaware and Hudson Canal, made six pictures in Cuddebackville, including *In Old Kentucky* (1909), starring Mary Pickford; *The Broken Doll* (1910), starring Dark Cloud (one of the only Native Americans to play Native American roles in the silent era); and *The Squaw's Love* (1911), starring Mabel Normand.

Further upstate, "Caribou Bill" Cooper formed the Arctic Film Company of Saranac Lake in 1909. He built a little log cabin village and a primitive one-room log cabin stage. The stage was patterned after Edison's Black Maria (America's first film studio), and equipped with a series of ropes and pulleys to regulate proper amounts of light by maneuvering the one muslin wall. Cooper rented out the facility, including a team of huskies in the bargain, to companies of actors from the Republic Motion Picture Company, Vitagraph, and Lubin for shooting northern-subject photoplays during the snow season and James Fenimore Cooper stories during the lush summer months. At other times he produced his own films. Full prints of the movies made at the Cooper camp seem to have disappeared, but the American Film Institute has a fascinating collection of still photographs documenting the work done there over a four-year period.

The American movie industry began its relocation to California in earnest after 1912. The outbreak of World War I in 1914 saw the curtailing of French, German, and English production. During the three-year period of American neutrality, the U.S. film industry thrived in its spacious new studios and excess of available sunlight and began to dominate worldwide film exhibition. Feature-length films of ninety minutes or more became common. Title cards made the longer films easy to follow, and new projectors, such as the Simplex, made longer films easier to watch by reducing visual stuttering and strobing. Feature films became the centerpieces of complete entertainment programs that often included shorts, serials, and live vaudeville acts.

3

☀ The Roaring Twenties ☀

Movie Palaces and Theater Organs

Several large theaters intended for mixed presentations of vaudeville and movies were built in New York City before the end of World War I, but the 1920s was truly the age of the movie palace. Several factors contributed to this, not the least of which was the general availability of cheap credit during the 1920s, which led to a national boom in many types of construction and investment (and eventually, according to most economists, to the Great Depression). The emergence of large vertically integrated studios in the American movie industry was another factor. As films became longer and more expensive to produce, a shakeout occurred among the many production houses that had set up shop in early Hollywood. After the smoke of bankruptcies, mergers, and acquisitions had cleared, the survivors had consolidated into seven large studios that ruled Hollywood and much of the international film world until the television era. It was a two-tiered group consisting of the majors—MGM, Paramount, Warner Bros., and RKO—which had their own national theater chains, and United Artists, Columbia, and Universal, which did not. Theater ownership allowed a studio complete control of a film property from concept to ticket booth. Studios that could assure their own distribution markets functioned as manufacturer, wholesaler, and retailer of their products, sharing no profits with middlemen.

B. F. Keith's, founded as a chain or "circuit" of vaudeville theaters in the 1880s, was one of the first such companies to recognize the movies as the wave of the future. In 1896, Keith's booked the Lumière Brothers into its Union Square Theatre in Manhattan for the first American demonstration of its film projection system. Impressed by the crowd's reaction, Keith's made a national distribution deal with Biograph a few weeks later that lasted for a decade, ending when Edison offered a better deal. As a twenty-five-year veteran of the movie business (which was only about twenty-five years old), Keith's was well positioned for the expansion into large theaters that took place in the 1920s.

The 1920 opening of B. F. Keith's at 408 South Salina Street (see plates 4–6) ushered in the era of the movie palace in downtown Syracuse. The theater opened its doors with two-a-day reserved seat presentations of a mixed vaudeville-movie

37. RKO Keith's was called the most beautiful theater in the country when it was built in the 1920s. It was torn down to make way for Sibley's department store. Its organ is now at the New York State Fairgrounds. Courtesy of Barbara Thomson.

show. The Keith circuit brought the great stars of vaudeville to Syracuse, many of them recently made even more popular by radio: comedians Jack Benny, Eddie Cantor, Milton Berle, and George Burns and Gracie Allen; magicians Harry Houdini and Harry Blackstone; and singers Sophie Tucker and Al Jolson; as well as headliners offering every kind of popular entertainment: ventriloquists, dancers, even trained animals. The theater itself was part of the show. The Keith had elegant marble pillars, rich silk hangings, murals on its walls, and a broad staircase leading up to the expensive loge and cheaper balcony seats. The stage boasted fifty thousand dollars' worth of theatrical lighting and sound amplification, and each of its twenty-five dressing rooms was equipped with a shower and tasteful furniture. In 1928, Keith's partnered with the Radio Corporation of America—the most powerful communications in the world at that time—to

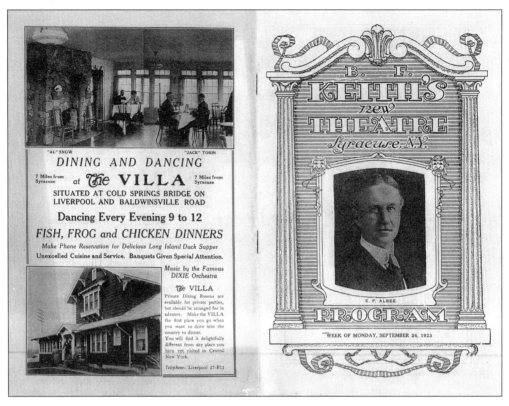

38. Program for B. F. Keith's Theatre, Syracuse, New York.

launch a new movie studio, RKO (Radio-Keith-Orpheum). From then on, each theater in the circuit was known as the RKO Keith. The RKO Keith on Salina Street was demolished in 1967 to make way for Sibley's department store. Its twenty-five-thousand-dollar Wurlitzer organ continues to entertain audiences at the New York State Fairgrounds in Syracuse, where it is played in monthly recitals under the auspices of the Syracuse Theatre Organ Society.

Grand theaters were also to be found in the neighborhoods. In fact, the Palace Theatre at 2384 James Street in Syracuse's Eastwood section, complete with a Wurlitzer organ as well, was built by Alfred DiBella in 1924. Its quieter location, away from valuable downtown real estate, may have saved the Palace from the wrecker's ball, which was the fate of the RKO Keith and other downtown competitors. The Palace has remained in continuous daily operation since it opened, going dark only for renovations. For nearly fifty years, it was run singlehandedly by Francis DiBella, Alfred's daughter. Upon Francis's death in her eighties, her nephew, Michael Heagerty took over the theater. He completely rehabilitated it, restoring it to original proportions and adding a new marquee, lighting, carpets, and other features, including a contemporary projection booth.

39. B. F. Keith's vaudeville theater, which later became RKO Keith's, Syracuse. Courtesy of Barbara Thomson.

In 1926, the Harvard Theatre opened on Westcott Street, at the center of the neighborhood adjacent to the east side of the Syracuse University campus. Pseudo-Spanish in style, its interior was red, blue, and gold, and it had indirect ceiling lighting as well as direct lighting from eight chandeliers and sidelights. In keeping with the Spanish theme, the exterior of the building had tiles and white- and red-spiraled columns. As any respectable movie theater built in the 1920s, it was equipped with a Wurlitzer. During the 1930s and 1940s, Professor Sawyer Falk of the Syracuse University drama department used it for his Boar's Head Theatre stage productions. Renamed the Westcott, the street-access, single-screen theater remained a rare asset—an independent neighborhood theater—until October 2007 (see plate 7). Along with the Manlius Art Theatre, also a neighborhood indie, the Westcott was operated by Nat Tobin, who usually said a few words to the audience about the evening's presentation

40. The Acme Theatre, on the corner of Butternut and Park streets, was owned by S. P. Slotnick. This picture is undoubtedly of a matinee performance sometime in 1938. Collection of the Onondaga Historical Association, Syracuse, New York.

41. The Palace Theatre on James Street in Eastwood was built by Alfred DiBella in 1924, two years before Eastwood became part of Syracuse. It is still in operation, run as a single-screen theater by Alfred's grandson, Michael Heagerty, who completely refurbished the theater with state-of-the-art equipment and a coffee shop in 2005. Collection of the Onondaga Historical Association, Syracuse, New York.

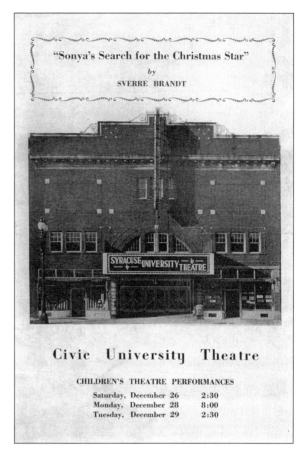

42. This Civic Theatre program for "Sonya's Search for the Christmas Star" reflects Syracuse University's use of the theater for its shows in the 1930s.

before the lights went down. Although the Manlius is still operating, financial reasons forced Tobin to close the Westcott in 2007, to much local dismay.

In 1929, two grand theaters were built on the south side of the city. The Brighton, with 1,750 seats (see plate 8), stood at 2616 South Salina Street, and the Riviera Theatre (with 1,100 seats), at 3120 South Salina. The latter, gone since a 1968 snowstorm caused its ceiling to collapse, sported Moorish interior decor and a star-studded ceiling.

Back downtown, the Paramount Theatre, boasting fifteen hundred seats, was built in 1929 by the Publix Theatres Corporation. It stood at 424 South Salina Street, the site of the Temple Theatre (built in 1914), which was gutted to make way for it. Also in 1929, the jewel in the crown of all Syracuse movie houses opened: Loew's State Theatre at 362 South Salina Street (see plates 9–11). The largest cinema in town with 2,900 seats, the State was designed by Thomas White Lamb, architect of many of the great Loew's movie palaces. The Syracuse project was said to be Lamb's favorite, and he used it as a prototype for two Manhattan theaters, the Loew's 72nd Street on the Upper East Side (demolished in 1961) and the Loew's 175th Street in Washington Heights (now a church).

43. The Brighton Theatre at 2616 South Salina Street was a large neighborhood house. It later became a bowling alley and now houses the Dunk and Bright furniture store. Collection of the Onondaga Historical Association, Syracuse, New York.

44. The Riviera Theatre, located at 3120 South Salina Street in Syracuse, had a Moorish decor with a star-studded ceiling. During the 1960s it showed art movies, but the roof caved in from excessive snow in 1968. Courtesy of Barbara Thomson.

45. A crowd gathers outside the Riviera Theatre in Syracuse, New York, on opening night. Courtesy of Allen Kosoff.

46. Riviera Theatre interior. Courtesy of Allen Kosoff.

47. The Paramount, at 424 South Salina Street, was the successor to the Temple Theatre. It was torn down to make way for Sibley's department store. Courtesy of Barbara Thomson.

The Loew's State in Syracuse . . . Imagine what it was like to enter such an emporium of dreams! A liveried doorman welcomes you into a grand lobby decorated in Persian-Byzantine style. A gorgeous Tiffany chandelier, something out of the old Vanderbilt mansion in New York City, lights the room. Gigantic pillars, bas reliefs, and enormous murals frame the space. A grand carpeted stairway leads up to a mezzanine, furnished with antiques from around the world. You were transported to another world before even entering the auditorium. If a patron felt like a princess every time she walked down that grand stairway, then the Loew's decorators had done their job.

A large staff of uniformed doormen, ushers, restroom attendants, and coat-check clerks stands at the ready. An usher escorts you to your seat. (Many Syracuse University students earned their way through school ushering at the State, the Paramount, or another of the downtown movie palaces.) You enter the enormous auditorium. The Mighty Wurlitzer is filling it with sound. The enormous stage is

48. Loew's State Theatre was the gem of Syracuse's movie theaters. Saved from threatened destruction in the 1970s, it remains in operation today as the Landmark Theatre. Courtesy of Barbara Thomson.

49. Interior of the Loew's State Theatre (now the Landmark Theatre) in its heyday. Courtesy of Barbara Thomson.

50. Syracuse's "Broadway Row" on South Salina Street in the early 1930s. Courtesy of Barbara Thomson.

bathed in multicolored lights below a great domed ceiling. Thomas White Lamb expressed the philosophy behind his movie palace designs: "The theatre is the palace of the average man. As long as he is there, it is his, and it helps him to lift himself out of his daily drudgery." Marcus Loew, head of the company, put it more simply: "We sell theaters, not movies!"

The program typically began with an overture, either by the orchestra or the organ (they alternated from show to show). Then there was a short film, either a comedy or a travelogue. A live musical number, either vocal or instrumental, would follow. The State, like many of the top-of-the-line palaces, had its own men's chorus. Then, perhaps, a vaudeville act or a short ballet number was presented, followed by either a cartoon or a newsreel and, finally, it was time for the main attraction!

Contrast that experience with the typical contemporary movie-going experience. You travel to a mall and find your way through a sea of cars. You enter a glitzy lobby dominated by an enormous refreshment stand, usually running its full length. Making your way past it, you enter a long, dark, narrow corridor punctuated by a series of nondescript little doors leading to separate screening rooms, some with as few as two hundred seats and screens not much bigger than widescreen televisions. This theater is utilitarian at best: no distinguishing decor, no special lighting, nothing to stimulate the imagination—only junk food to overstimulate

51. The System Theatre stood in the 500 block of South Salina Street. It was also called the Top, the Syracuse, the Ritz, and later the Civic. Collection of the Onondaga County Historical Association, Syracuse, New York.

the appetite and overcharge the customer. As you wait for the film, Muzak emanates from the very advanced sound system, as a slide show tries to convince you to go back to the lobby to buy candy, popcorn, and soda. Finally, the program begins—too many coming attractions at too high a decibel level offer too many special effects that are meant to impress you, rather than to express human emotion to you. At last! The feature film—and that's it. Time to go before they kick you out. No shorts, no cartoons, no newsreel, no second feature! Price? Ten bucks! What is wrong with this picture?

The Theater Organ: Instrument of Choice at the Movie Palaces *

Silent films were far from silent. They were almost always shown with some kind of musical accompaniment. Depending on the size of the city and the theater, music might be supplied by a single simple upright piano, a full symphony orchestra, or any of the configurations of musicians that lay in between. The more important films—features by D. W. Griffith or Charlie Chaplin, to name two prominent examples—had fully composed musical scores. But most films arrived at the theater accompanied only by a cue sheet, a set of general instructions giving the pianist a hint of when moods shifted and what type of themes might be appropriate. Piano accompaniment was a challenging art requiring special talents, and many accomplished musicians were unable to perform at the movies. You had to be mentally and physically nimble, with a dozen or more themes at your fingertips. Theme

* Written with assistance from Will Headlee and Stu Green.

books were published offering suggestions of tunes for every mood. But a large repertoire was not enough. A successful film accompanist had a jazz musician's ability to improvise segues that mirrored the rhythm of a film's cutting and emotional pitch. Sudden changes, natural progressions, incongruous juxtapositions, and even rhetorical styles (such threat or sarcasm) had to be woven together into a seamless whole.

The theater organ was a new kind of musical instrument created especially for these purposes, and it came to the forefront of public attention during the 1920s, as impressive new theaters, intended exclusively or principally for the exhibition of films, proliferated. Theater organs differed from church organs in that they had a number of extra theatrical stops simulating sound effects (e.g., horse hooves, screams, shattering glass) and simulating other musical instruments (e.g., horns, drums, bells). At the larger theaters, the organ often alternated shows with a full orchestra, playing an overture and interludes as well as simultaneous film accompaniment. Although the theater organ was widely enjoyed by audiences and intensely loved by aficionados, its heyday was shortened by a stunning technological development. In 1927, Warner Brothers released *The Jazz Singer*, the first commercially successful "talkie" (that is, a film with an automated synchronous soundtrack). Within two years, virtually all nontalkie feature production had ceased and the specially developed theater organ became obsolete.

Nonetheless, audiences continue to enjoy theater organ music on its own merits. Radio City Music Hall in Manhattan, which has the largest pipe organ built for a movie theater, is synonymous with the instrument in the minds of many people today. Installed for the 5,940-seat theater's 1932 opening, the Radio City Wurlitzer was not built to accompany silent movies, but rather as a concert instrument. Tens of thousands listen to it played each year at the variety of live and film events that are booked into Radio City.

Another surviving downstate Wurlitzer is still at work in the gymnasium of Long Island University's Brooklyn Center, which was once the Brooklyn Paramount Theatre, built in 1928. Converted to a collegiate sports facility in the mid-1960s, the building retained the organ, which is played regularly during LIU intercollegiate basketball games and occasionally in university-sponsored concerts.

In upstate New York, where Wurlitzer and other manufacturers made most of the best theater organs of the movie palace era, there is a great deal of interest in theater organ music, with several concert venues for devotees of the instrument. Binghamton offers numerous and varied opportunities. The Forum Theatre, a performing arts center originally built as a movie house on Washington Street in downtown Binghamton, boasts a 1924 theater organ manufactured by the Robert Morton Pipe Organ Company of Van Nuys, California, originally installed at the America Theatre in Denver. The Binghamton Theater Organ Society sponsors

52. Wurlitzer organ. Courtesy of the Empire State Theater and Musical Instrument Museum.

a concert series at the Forum each year. For outdoor listening, two carousels in Binghamton city parks have theater organs in daily operation from May through October: Ross Park Carousel has a Wurlitzer 146-A band organ, and Recreation Park Carousel has a Wurlitzer 146-B. Binghamton's Roberson Museum and Science Center owns not only a theater organ but also several player pianos and automatic instruments, all specialties of the city's Link Piano Company.

Ithaca's State Theatre, like the Forum, is a performing arts center that began life as a movie palace. Its original theater organ is a Link, built in Binghamton in 1926, which it acquired in 1928, second hand, for twenty-six thousand dollars from a traveling vaudeville show that went broke. The State opened its doors to the public on December 6 that year, bringing in Harry Springer, a name accompanist from Chicago, to play the overtures and scores to a double feature of *West of Zanzibar* starring Lon Chaney and *Show Girl* starring Alice White. The Link organ was played daily at the State for decades until high-fidelity sound and changing movie industry economics turned it into a dinosaur. In 1968, Robert Engel, an organ restorer from Syracuse, bought the decaying instrument and removed it, piece by piece, from the theater (saving it from destruction in the process). After the turn of the twenty-first century, the State was converted from

53. The State Theatre in Ithaca is currently undergoing renovation as a community center.

a moribund movie theater into a performing arts venue by Historic Ithaca, a non-profit organization. A committee was formed, which located, bought, restored, and reinstalled the organ at the State. It resumed entertaining Ithaca audiences on December 7, 2004, exactly seventy-six years and a day after its first Ithaca premiere.

In Syracuse, a 1925 Opus 1143 Wurlitzer Unit Orchestra, originally installed at the Keith's on Salina Street, is used for monthly concerts at the Empire State Theater and the Musical Instrument Museum at the New York State Fairgrounds, which does business as the Empire State Expo Center when the fair is not running. This is the last known surviving theater organ of the eighteen that once played in movie theaters in every quarter of the city.

New York State Theater Organ Manufacturers

Wurlitzer, made in North Tonawanda, a suburb of Buffalo, was and remains the king of theater organ manufacturers. The company dominated the market so thoroughly that for a time all theater organs, no matter the brand, were known generically as "mighty Wurlitzers." During the heyday of the movie palace, Syracuse could boast eight Wurlitzers among the eighteen theaters in the city equipped with organs.

CONSOLE OF THE FIVE-MANUAL

Marr and Colton Organ

INSTALLED IN THE

ROCHESTER THEATRE

ROCHESTER, N. Y.

This organ is erected on an elevator platform and is raised or lowered by push-button control, operated by organist. This five-manual organ is one of the largest organs ever built. The organist will find nothing lacking. He can get whatever tonal effects, ample in volume and quality, whenever he wants them—easily and quickly.

The Marr & Colton Co. (1915-1932) produced theatre

organs in the Wyoming Co. shiretown of Warsaw, N, Y.

Courtesy of the Warsaw (N.Y.) Historical Society.

54. The Marr and Colton Company (1915–32) produced theater organs in the Wyoming County shire town of Warsaw, New York. Courtesy of the Warsaw (N.Y.) Historical Society.

For Syracuse theater owners looking for something a little less expensive, a likely choice was a Marr and Colton, manufactured in Warsaw, New York, southwest of Rochester in Wyoming County. With a Marr and Colton, as many theater organ aficionados are apt to say, you never knew quite what you were getting. They were uneven in quality and often had parts that made weird or unexpected sounds.

A more reliable, if more expensive, choice was another upstate manufacturer, the Link Piano Company of Binghamton, which was known mainly for its pianos and its automatic instruments, including traditional player pianos and "orchestrions," which were machines that imitated entire orchestras. Link manufactured approximately one hundred theater organs between 1925 and 1929. After the stock market crash, Edwin Link, Sr., the company's founder, retired. His sons

Edwin, Jr., and George took charge of the company, redirecting it toward making flight simulators used for training pilots. Other theater organ manufacturers located in upstate cities included Barnes and Buhl (Utica), Beman (Binghamton), Carey (Troy), Delaware (Tonawanda), Hope-Jones (Elmira), Kohl (Rochester), and Schickler (Buffalo). Downstate organ producers included Midmer-Mosh (Merrick, Long Island) and Roosevelt (New York City).

In downtown Syracuse, The Strand, Regent, and Eckel theaters were all originally equipped with small, colorless Austin organs. The Strand replaced its Austin with a Style-240 Wurlitzer, circa 1926, and the Eckel switched to an eight-rank Wurlitzer at about the same time. Many downtown houses, including the Empire, Crescent, Rivoli, Bastable, Temple, Syracuse, and B. F. Keith's, as well as the Wieting Opera House, used pianos for silent film accompaniment or hired a small ensemble from the musicians' union local. But after the Rivoli, a modest, second-run house near Syracuse's huge New York Central Railroad station, installed a two-manual-console–six-rank (2/6) Marr and Colton, organ theaters became *de rigueur* in the city. The Rivoli's managers, the Fitzer Brothers, made the most of a grand opening for the organ, and other theater owners took note of the long lines waiting to get into the Rivoli. Within a year, most downtown exhibitors had organs or had set dates to get them, clearing spaces for chambers, wind lines, and blowers. The large consoles often took up areas that had been occupied by three musicians in the already cramped orchestra pits. The Novelty, across the street from the Rivoli, was a remnant of the early converted storefront movie houses that had specialized in blood-and-thunder films. Its flat front was a garish billboard of colorful lithographed posters advertising the current attraction. It was equipped with a Robert Morton FotoPlayer theater organ, consisting of a few short ranks of pipes, supplemented by a piano and several sets of free reeds in two swell boxes.

The magnificent Loew's State Theatre did not open until 1929, two years after the premiere of *The Jazz Singer*, the first successful talking picture, which had doomed silent film production. The State was nevertheless equipped with a very good 4/20 Wurlitzer. But none of the theater organs in Syracuse or in the Central New York region ever quite reached the tonal excellence of the 3/11 Wurlitzer at the Keith. One of the finest such instruments made, it now resides in the New Times Theatre at the New York State Fairground.

The Savoy on Warren Street had a piano with several ranks of pipes available on an upper 61-note manual, and a one octave pedal division was added in the early 1920s. The organist, Luella Edwards Wickham was overjoyed by the addition of the pedal division because it enabled her to play overtures.

The Avon Theatre on Hawley Avenue, which opened in the 1920s in a largely Italian neighborhood, was equipped with a very sweet 3/8 Marr and Colton. The owner was also the organist, and he had a reputation for a greater interest in wowing the ladies than in excelling in business or music. A devotee of celebrated theater organist Jesse Crawford, he saw himself as a soloist and put much effort into his sing-alongs.

During the mid-1920s, the city fathers expected Syracuse to expand down South Salina Street. Builders were encouraged to erect downtown-style office buildings, stores, and theaters along the main thoroughfare in the typical streetcar-based expansion pattern of U.S. cities during this period. The expected southward thrust failed to materialize in any volume, but efforts resulted in the construction of some impressive isolated buildings. One of these was the Brighton block, which housed a large theater with a fine 3/10 Marr and Colton. Unfortunately, the organist could not hear music played on the theater organ clearly until it bounced off the back wall of the auditorium; the chambers were aimed straight out from the balcony level. The Brighton Theatre, built on a grand scale, was several miles from downtown and, despite vaudeville acts, a pit orchestra, organ soloists, first-run movies, and a much-ballyhooed start, it soon went dark. The theater was first converted into a bowling alley. Eventually, it became the large Dunk and Bright furniture store.

The Brighton fiasco does not tell the entire story of South Side theaters in Syracuse. Smaller, less ambitious houses prospered in the same general area. The Arcadia, at the corner of East Colvin and South Salina streets, was the proverbial hole-in-the-wall movie house. It opened long before the Brighton did and continued operating long after the Brighton closed. A second-run house, the Arcadia did well offering movies suitable for children on Saturday afternoons. It had a roll-playing Seeburg piano with two ranks of pipes (stopped flute and short quintadena) that were played from a manual above the keyboard. Pipework was contained in the oversize piano case; the low flute pipes, which went down to bass C, were attached to the back of the piano case horizontally where they played, without expression. Action was pneumatic. Tone quality was on the harsh side. The only added effects were guitar buttons, which a stop-key would cause to lay against some of the piano strings for a plucked string effect. As long as the Arcadia ran silent movies, no musician's fingers graced the instrument's keyboards. It was used exclusively to play rolls that occasionally provided music to fit the action on the screen. The customers didn't seem to mind, and the Arcadia operated well into the talkie era using hand-me-down equipment that major houses considered obsolete.

One of the more interesting neighborhood houses was the Riviera, at 3116 South Salina Street, near the corner of West Newell Street. When Harry Gilbert sold the Regent in 1928, he used the money to realize his dream theater, to be built in a nice neighborhood for family entertainment. It was an atmospheric house, built to resemble an Italian garden with a twilight-blue ceiling, complete with cloud projector and twinkling stars. For a rather tiny theater, the organ chambers were huge, far larger than required by the 2/7 Wurlitzer. Gilbert had made a deal for a much larger organ of another make that had fallen through, so he settled for the small Wurlitzer. The Riviera opened as a talkie house, with only a few silent films booked for the kids on the weekend. Actually, it didn't require an organ at all. Yet the showman in Gilbert told him that his customers had so long connected organ music with a trip to the movie theater that they would expect it, whether it was needed or not. Gilbert wanted everything about his new Riviera to be first class, and in 1928–29, with the 2/7 Wurlitzer in the house, he hired one of Syracuse's best musicians, Bart Wright, to play song slide novelties, overtures, film preludes, and intermissions.

The Harvard (later the Westcott Cinema), at the corner of Harvard Place and Westcott Street near Syracuse University, opened in 1927 with a Style B Wurlitzer whose four ill-matched ranks might have served as the "brain" (accompaniment) for a larger instrument. The one good voice was a Style D trumpet. The salicional was coarse in tone; the concert flute nondescript and dull; the vox croaked; and worst of all, the single chest action was noisy so that a continuous clacking percussion was heard when the instrument was played.

A much better instrument was the 2/3 Wicks across town in the Elmwood on South Avenue. It was also a roll player. Its stopped flute, vox, and string were matched in a pleasing tonal ensemble. Organist Vic Viveros made it sound much larger than three ranks as he played blood-and-thunder silent films for the Saturday afternoon matinee.

The end of the 1920s brought a double whammy to movie theaters. The introduction of sound required new equipment, and outfitting a theater, even a small one, cost at least fifteen thousand dollars. But a Great Depression began after the stock market crashed in 1929, and that kind of capital was very difficult to come by. The combination of circumstances put the owners of many small- and medium-sized theaters out of business.

4

The Great Depression and the War

Talkies, Double Features, and Dish Night

The Jazz Singer (Warner Bros., 1927), the first commercially released feature with a synchronous sound track imprinted on the film, premiered in New York City in November 1927. The film was not a "talkie" but rather a silent-era film, complete with intertitles. The only words the audience was supposed to hear were the lyrics of the songs. The movie industry's original interest in sound film was not to make talkies, but rather to increase profits on the kinds of movies they had been making by cutting out the cost of hiring live musicians, an overhead expense that was attached to every screening at every movie theater. As an added benefit, producers could assert quality control over scoring by supplying a single professionally polished sound track for each film, no matter where it was exhibited. But the success of *The Jazz Singer*—and especially of Al Jolson's unscripted ad-libs—exceeded expectations, creating an instant demand for "all-talking" pictures. With the notable exception of Charlie Chaplin, who saw silent film as an art form—*his* art form—rather than a technologically backward craft, Hollywood remade its physical plant for the exclusive production of sound film.

Talkies were already a fact on the ground of the movie world in 1929 when the New York Stock Market crashed. More than a decade of economic hardship followed, including unemployment rates hovering at the 25 percent mark and underemployment and uncertainty in most of the work that was available. In short, the 1930s were a great time to go to the movies, if you could afford the price of a ticket.

Some film historians tend to spotlight the Warner Brothers studio and its high profile movie project starring Al Jolson, one of vaudeville's most popular performers, as the genesis of talking pictures. The more knowledgeable among them usually mention that sound film was not an in-house innovation at Warners but a technique developed by Western Electric, an engineering and manufacturing arm of the American Telephone and Telegraph Company (AT&T). Not much was said

55. Theodore Case, inventor of the first commercial
sound-on-film process, at Casowasco, his summer
home on Owasco Lake. Collection of the Case
Research Lab Museum.

about the company or the individuals who developed its sound film technique,
largely because the terms of AT&T's virtual monopoly on American telephone
service made Western Electric's involvement in nontelephone businesses legally
questionable. The telephone company was, in fact, barred from the movie business—but the court battle lasted until 1956. Meanwhile, AT&T did quite nicely.
In 1928–29, for example, Electrical Research Products, Inc., a subsidiary of AT&T's
Western Electric subsidiary, refitted more than 3,250 movie theaters in the United
States for sound film.

If AT&T was seeking a low profile for its contributions to the invention of
sound film, an upstate New York backyard inventor who beat Ma Bell to the
screen had obscurity thrust upon him. Theodore Case developed his sound-on-
film process at his home in Auburn, New York, about twenty miles west of Syracuse. Case, like several other brilliant engineers of the early twentieth century,

56. Case and his assistant, Earl Sponable, in their
laboratory in Auburn, New York. Collection of the
Case Research Lab Museum.

had the misfortune of working with Dr. Lee DeForest, who took much of the
credit for Case's work. But the record is clear. Case made the breakthrough and
filed the patent in 1923, selling the process to studio founder William Fox for use
in the company's Movietone newsreel division.

Adopting the slogan "Fox Movietone News: It Speaks For Itself," the studio premiered the system to the public in May 1927, with coverage of Charles Lindbergh's
departure from Roosevelt Field, Long Island, on the first successful transatlantic
solo flight. Shown packing five sandwiches aboard *The Spirit of St. Louis,* Lindbergh
was asked by the Movietone reporter if that was all he was taking. "If I get to Paris,
I won't need any more. And if I don't get to Paris, I won't need any more, either,"
he said. *The Jazz Singer* premiered four months later.

The Cayuga County Museum in Auburn kept Case's workshop intact for
decades, and now the Case Research Lab, as it is officially known, contains an
original Fox Movietone mobile unit newsreel truck, Unit 13, painted bright green

57. Interior of Case Research Lab, Auburn, New York. Collection of the Case Research Lab Museum.

with the company name and address lettered on its side: Fox Case Corporation, 460 West 54th Street, New York City (see plate 12).

Depression Talk

The 1930s represented America's first great crisis of confidence since the Civil War. Weary of economic struggle in a society that only recently transformed from a farming to an urban majority, people longed for reaffirmation of individual identity to restore a sense of individual potency. Many found this in the wave of gangster and private-eye films that packed the houses in the early 1930s. They discovered comic relief in the anarchism of the Marx Brothers and W. C. Fields and in such screwball romantic comedies as *The Awful Truth, Bringing Up Baby, Topper,* and *Nothing Sacred.* The rise of the independent woman was exemplified in a variety of screen personas created by such stars as Jean Harlow, Mae West, Carole Lombard, Katharine Hepburn, Myrna Loy, and Jean Arthur. As the decade wore on, a yearning for law and order was addressed in the advent of G-man films whose heroes were federal crime experts, and a revival of Westerns, whose heroes had been admired since the nation's founding for their self-reliance and innate sense of justice. In the 1930s film seemed to replace the novels and short stories that had so moved the young intellectuals of the 1920s. The movies

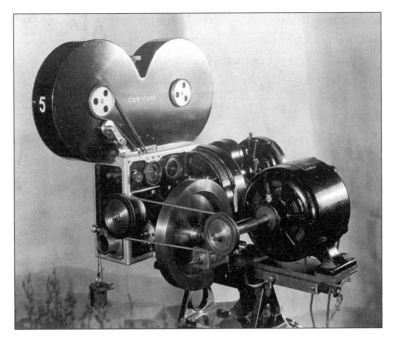

58. Fox-Case camera using Case sound equipment. Collection of the Case Research Lab Museum.

were supplying heroes and shaping dreams. The director Federico Fellini, who grew up in Mussolini's Italy cherishing any opportunity to see a Hollywood film, put it this way: "The movies and America are almost the same thing."

Times being what they were, theaters had to come up with gimmicks to entice audiences to spend what little leisure money they had on the movies. Double features—two pictures for the price of one—were a popular innovation, as were dish nights, which supplied weekly moviegoers with an entire set in just a few months. They tried every kind of promotion: bingo, raffles, talent contests, and sing-alongs ("Follow the bouncing ball!"). For the kids, there were Saturday matinees with secret rings, playing cards, candy, and, of course, on the screens, there were the serials, newsreels, travel shorts, and cartoons. So-called "B" studios, such as Mascot, Chesterfield, and Republic, arose, producing low-budget films for the bottom half of double bills. Theaters introduced air conditioning during the thirties. Since it was not available residentially, it improved attendance at most theaters during the summer. Moreover, it allowed theaters in the hottest sections of the country, which formerly had had to close during the summer, to stay open year-round. Since air conditioning plays a role in movie history, it is worth mentioning that the process was invented by Willis Carrier, a Cornell graduate from Angola, New York, who also established the first plant for the mass manufacture of air conditioners in Syracuse. The 1930s also saw the first movie houses outside the segregated South to cater specifically to African Americans. New York City had more than a dozen such theaters and Buffalo had seven.

For a child growing up in central New York, a Saturday afternoon meant the movies: a nickel for the show, a penny apiece for candy, and an extra penny for a grab bag. Until the 1940s, theaters rarely sold candy. There was usually a candy store or soda parlor next to a theater, and in many upstate cities, they were run by Greek immigrants. Patrons bought candy before entering the theater, or had a soda or a sundae afterward. In Syracuse, the Schrafft's restaurant on Warren Street was a favorite rendezvous after the show. Oh, those hot fudge sundaes!

Neighborhood theaters usually rotated shows three times a week: a Sunday-Monday program; a Tuesday-Wednesday-Thursday program; and, of course, Friday and Saturday. It was not unusual for entire families to go to the movies several times each week. Gore Vidal hit the nail on the head when he said, "It occurs to me that the only thing I ever really liked to do was go to the movies." Arthur Schlesinger, Jr., echoes that thought: "I remember the movies of the 1930s and 1940s better than the movies I saw last week."

So many shows required the transportation of an awful lot of heavy 35 mm film cans. Syracuse was in the Buffalo district of film exhibition. All the major film companies had offices and warehouses in Buffalo. A trucking company would cart films in the middle of the night, bringing in new reels and taking out the old. A person could manage a theater for decades and never see the trucks that delivered the films.

The Great Depression did not end with a happy revival of American prosperity, but with the economic opportunities created by a frightening new war in Europe and the Pacific, beginning in 1939 with the German invasion of Poland. As bad a year as that was for humanity, it was a banner year for movies, with such films coming out of Hollywood as Gone with the Wind, Goodbye Mr. Chips, Wuthering Heights, Stagecoach, Mr. Smith Goes to Washington, and The Wizard of Oz.

As the 1930s came to a close, color emerged as the next great technological frontier in cinema. Actually, color had been used since the earliest days of the medium, when films were hand-tinted by stencil. In Birth of a Nation, D. W. Griffith used the French practice of tinting scenes for dramatic effect: blue for night, orange or yellow for a sunny day; red for violence, and so on. During the twenties, color sequences were still considered experimental and were very expensive, but they were occasionally employed for effect in important scenes. Cecil B. de Mille used color sequences in The Ten Commandments (1923), including the scene in which Moses receives the tablet on Mount Sinai, and in Ben Hur (1925) for the chariot race. Douglas Fairbanks's The Black Pirate (1926) was shot entirely in a pioneering two-color Technicolor process. Walt Disney introduced a three-color process in *Trees and Flowers* (1934), an animated cartoon. The first feature to be shot entirely in three-color Technicolor was *Becky Sharp* (1935), starring Miriam Hopkins.

The two great films that made pioneering use of color were both released in 1939. *The Wizard of Oz* used sepia for the "realistic" Kansas scenes at the film's opening and conclusion. *Gone with the Wind* was probably the ultimate Technicolor picture made to that point in time. But Technicolor was expensive and difficult to shoot. Wartime prohibitions on the use of chemicals suspended all color production from the time of the U.S. entry into World War II for the duration. When color shooting resumed in the late forties, a multitude of cheaper color processes appeared, but the color soon faded, leaving only garish red and orange tones. Many of the prints of these films are now unviewable and many early color films reissued for television are inferior and perhaps should not be shown. During the late eighties and early nineties, Turner Television, which owns several important film libraries, began colorizing some of the old black-and-white classics, to the consternation of film purists. The brilliant sheen of old black-and-white prints simply cannot be duplicated in color. The purists claim that even now, some films should be shot in black and white. But the television industry believes that television audiences demand color.

The War Years

For the theaters that survived the Great Depression, the early 1940s proved a bonanza. War industries put money into people's pockets, but they had few places to spend it. Food and clothing were rationed, as were gas and tires. Big-ticket items, including automobiles and refrigerators, were not being produced for the consumer market. The military was given priority on the passenger railroads, and civil aviation, which had barely gotten off the ground during the thirties, was suspended. Radio was very much in its heyday, but there wasn't much to do outside the house. Professional and college sports were in the doldrums, with many of the best players away in the service. It got to the point where Joe Nuxhall, a sixteen-year-old, was pitching for the Cincinnati Reds. The only thing left to do was to go to the movies—and people went in record numbers. Theaters were open around the clock in areas where swing-shift workers were plentiful. While some male stars served in the military, others were encouraged to continue to make movies and, between productions, to perform for the troops in shows organized by the USO (United Service Organizations). There were many popular female stars and plenty of upcoming young leading men to play opposite them.

Some films made during the war were unabashed propaganda in which heroes killed every "Jap" and Nazi in sight. Others, somewhat subtler, were the result of ongoing cooperation between the military and the studios over what messages ought to be communicated about the war. But propaganda was not a big seller. The most popular wartime fare was escapist farce designed to take the viewer's thoughts away from the horrors it caused for those who fought and the daily inconveniences it created on the home front. Darker reflections would come in pictures made after the war.

Comedy was king. The broadly played "Road" pictures, starring Bob Hope, Bing Crosby, Dorothy Lamour, and Jerry Colonna, epitomized wartime Hollywood cinema, as did the more sophisticated comedies directed by Preston Sturgis: *The Great McGinty, Christmas in July, Sullivan's Travels, The Lady Eve, The Palm Beach Story, The Great Moment, The Miracle of Morgan's Creek,* and *Hail the Conquering Hero.* The Marx Brothers, Jimmy Durante, and Abbott and Costello all did well. Lighthearted Latin-themed musical comedies, with stars such as Carmen Miranda and Xavier Cugat, suddenly proliferated as the studios reached out to Latin American markets to make up for the loss of overseas audiences in Nazi-ruled Europe and Japanese-ruled East Asia.

Stars went from city to city raising money for the war effort at bond rallies, and from battlefront to battlefront attempting to raise the morale of the troops. War Bond stamps, which could be saved and traded for war bonds, were sold at every theater, and free admission was granted in return for the donation of a pot or a pan to a scrap drive. On a good night, a scrap drive could turn an elegant movie-palace lobby into something resembling a junkyard. Newsreels were popular to audiences in which everyone knew someone exposed to the dangers of war. Major directors pitched in to help. Frank Capra, who joined the Army Signal Corps, made the "Why We Fight" series, an even less subtle than usual expression of his deeply felt patriotism. He also made military training films that were so entertaining they were shown in theaters to ticket-buying customers.

The mood of the country during the war was somber, noticeably different from the enthusiasm and idealism that had accompanied American entry into the First World War. We had a job to do and we had better do it so that we can bring the boys home again. Social commentators have looked back upon these years as a crucial period when the country came together for a shared purpose, following the class and ideological divisiveness of the Depression years. Tom Brokaw calls the World War II generation "the greatest generation," praising it for making the sacrifices necessary to save democracy in America and elsewhere in the world. We fought because we knew we had to. It was grim and agonizing, but it promised a better tomorrow—and the consequences of failure could not be considered as an option.

5

❧ End of the Studio Era ❧

TV, Widescreen Projection, and Drive-ins

Theater attendance reached a peak in 1947—and then the bottom dropped out. Today, many people think that the rise of television caused these box office woes for the film industry. A closer look at the historical facts says otherwise. As of 1948, the Federal Communications Commission had issued 108 television licenses, all of them for stations in large cities concentrated in three geographical clusters: the Northeast corridor; the Great Lakes rim; and the West Coast. Because many of these cities had multiple stations (New York City and Los Angeles, for example, each had six), fewer than thirty metropolitan markets had television service, and in those that did, it took years for significant numbers of consumers to buy television sets. More cities were clamoring for television, but in 1948, the FCC imposed a freeze on new station licenses to solve technical and compatibility problems. New TV station licenses were not granted until 1952, and it took at least a year for news stations to build their transmission towers and get on the air. By 1953, the year television began to grow into a truly nation phenomenon, the American film industry was already experiencing steep economic decline.

The root of the problem for the movies was legal. In deciding the case of United States v. Paramount in 1948, the U.S. Supreme Court decreed that the major film studios, by controlling production, distribution, and exhibition of their films, were corporations acting in restraint of trade. To remedy this, each studio had to exit at least one of these three areas of the movie business. MGM, Paramount, Twentieth Century-Fox, Warner Brothers, and RKO each decided that the easiest course of action was to sell off its theater chain.

Thus began the decline and fall of the studio system. With their own theater chains, the studios had been assured of ready markets for fifty feature films a year, plus second features (B-pics), shorts, serials, cartoons, and newsreels. Without these assured theater engagements, production and distribution could not be rationalized. The studios' first move was to eliminate B-pics and short films, the lifeblood of neighborhood theaters. Unable to exist on second runs of major films that had already played downtown, neighborhood theaters began disappearing in the early fifties. The studios also cut back the number of feature films they were making so that directors and actors were no longer assured of five and six productions a year. Many tried setting up their own companies, but few succeeded.

The Paramount antitrust decision was not the only cloud hanging over Hollywood during the decade following World War II. There were other reasons why it was not a good time to be making movies in America. The postwar era was one of disillusion. The factors leading to disaffection at the moment of victory in war were numerous: the difficult economic readjustments for returning veterans and home-front workers (especially women who were unceremoniously fired from jobs they had done well), the knowledge of the atrocities of the Holocaust, the commencement of the Cold War and the nuclear arms race. One of the most depressing phenomena of the postwar period for filmmaking and for other creative industries in America was the sudden and ferocious red-baiting that took place in the investigations of the House Un-American Activities Committee and the hearings held by Senator Joseph McCarthy. Some of Hollywood's best writers, actors, and craftspeople were blacklisted by the studios, losing their livelihoods in retaliation for exercising their constitutional right to freedom of expression. Moreover, the spectacle had the effect of creating a high degree of self-censorship, preventing those who were making films from freely expressing their opinions. This was making the movies duller, just as they needed to become more exciting than ever.

The Coming of Television

Up until the end of the Second World War, the broadcasting and movie industries had complementary interests and good relations. The studios promoted their films and stars on radio, and broadcasters were generally pleased with the ratings boosts that Hollywood celebrities gave to their programming. There were some attempts at minding each other's business. RKO, for example, was partially owned by the Radio Corporation of America (RCA), parent company of the NBC radio network. RKO had been established in 1929, two years after the Western Electric (AT&T) sound-on-film system had become standard in the industry. RCA chief David Sarnoff had a sound-on-film system of his own and decided that if the studios would not buy his equipment, he would establish a studio to prove its superiority. Sarnoff's partner in the venture was Joseph P. Kennedy, who brought several small studios and the Keith theater chain to the deal, all of which became the nucleus of RKO. By 1948, Sarnoff, convinced that television was about to send the entire movie industry the way of the horse and buggy, sold the studio to Howard Hughes. On the other side, Paramount Pictures had obtained a license for an experimental television station in Chicago in 1938. Deciding against going into the broadcasting business because it was government regulated, the company switched strategies after World War II and invested, in partnership with the Zenith Corporation, in a "closed circuit" television service it hoped to offer to tenants of the new luxury apartment houses that were being built on Chicago's Gold Coast during the postwar building boom. But with free TV stations going on the air, the public was not ready to pay for this primitive form of cable TV.

After the war, the twenty-year balance of power between the film and broadcasting companies was broken by a series of terrible blows to the movie companies. The antitrust ruling of 1948 forced them to divest their theater chains, just as it was becoming obvious that television was not some futuristic gadget or passing fancy, but a ready-to-go medium capable of delivering a movielike product right to the home of the viewer. Sarnoff let it be known that he thought both radio (the business he had personally dominated) and the movies would disappear with the advent of television. The studio executives circled the wagons for a war with the broadcasters. The broadcasters, led by NBC's Sarnoff and CBS chief William Paley, felt they could do without the movies. Both networks focused their dramatic efforts on theatrical rather than cinematic forms of entertainment. In drama, that meant "teleplays" (by such writers as Paddy Chayefsky and Rod Serling), which were heavy on dialogue and looked as if a camera had been placed on a tripod in front of a proscenium stage; in comedy, that meant variety shows (starring such comedians as Milton Berle and Jackie Gleason), which were video approximations of vaudeville.

It was a short war that ended in lateral integration. In 1951, ABC, the weakest of the radio broadcasting companies that had gone into television, announced a merger with United Paramount Theaters (UPT), the old theater division of Paramount, which had become an independent company after the federal antitrust order of 1948. With FCC approval of the merger in 1953, Leonard Goldenson of UPT took the helm at ABC. Goldenson was the first film executive to head a broadcasting company; moreover, his fledgling television network was desperately in need of programming. Having stronger ties with the studio heads than with Sarnoff or Paley, Goldenson made a deal with Walt Disney for a new weekly series called *Disneyland,* and with Jack Warner for three new hour-long action series from Warner Brothers, thus turning the movie studios into television programming suppliers in 1955. When Sarnoff and Paley saw ABC come back from the brink of bankruptcy to make a dent in their ratings, they abandoned the teleplay strategy and followed Goldenson's lead. By the end of the twentieth century, the surviving studios and television networks were all units of multimedia conglomerates: Paramount and CBS (Viacom); Universal and NBC (General Electric); Disney and ABC (The Walt Disney Company); Warner Brothers, HBO, Turner Broadcasting, and others (Time-Warner); Twentieth Century-Fox and the FOX network (News Corporation); and so on.

The studios endured, but the American moviegoing experience would never be the same. Saturday matinee serials disappeared as the school-age youngsters who once filled the theaters stayed home in droves to watch television, which in some cases meant reruns of Republic Pictures cowboy serials starring Gene Autry or Columbia Pictures shorts starring the Three Stooges, which had been staples of a Saturday afternoon at the movies for years. Travelogues, newsreels, short subjects, and eventually everything but feature films and the coming attractions disappeared. Why come to the theater for things you already have at home?

Widescreen Projection

The studios did everything they could to make feature films bigger, better, and more interesting than what was offered on television. There were 3-D movies, which required theaters to distribute special cardboard eyeglasses. Some moviegoers liked the effect; others complained of headaches. Widescreen shooting processes, such as CinemaScope (Fox), VistaVision (Paramount), and Panavision (independent), were great for sweeping views of the West, but not great at all for interior scenes and close-ups, where they seemed to distort the ten-foot-tall faces of leading men and women just as they got into the romantic clinches. In any case, theaters had to make their screens wider and taller, often at the sacrifice of seats. Of all these processes, none asked more of theater owners or promised more to viewers than Cinerama.

Cinerama was based on cineorama, a multiprojector exhibition process (requiring eleven projectors) patented by Raoul Gromoin-Sanson in 1897. It was developed into a sound-film process called Vitarama, later renamed Cinerama, by Fred Waller, a native of Brooklyn, who did most of his work in Huntington, New York, on Long Island. Cinerama was introduced to the public in 1952 with *This Is Cinerama*, a film made specially to show off the process in a grand premiere at the Broadway Theater in Manhattan, attended by Governor Thomas E. Dewey. Many of those attending found it a thrilling experience, putting the viewer right inside the film. Others complained that the seams—the points at which the three projected strips composing the screen image met—were obvious and destroyed the illusion.

Cinerama demanded the complete renovation of existing theaters. Not only did the screen have to be big, but it had to be composed of special material and have a special sound system. These were extraordinary demands to make on exhibitors. Perhaps the biggest problem was that so few films were made with the process (and fewer still that were any good) that it was not worth the expense. At first, Cinerama films had to be shown with three 35 mm projectors, each filling the giant screen with one-third of the film image. In 1956, a special 70 mm projector, capable of doing the job itself, was introduced.

Cinerama arrived in Syracuse in 1958 when a new 75 x 25 foot screen was installed at Schine's Eckel Theatre on East Fayette Street. Syracuse University officials, led by Board of Trustees President Albert B. Merrill, held a dinner party in honor of the January 27 premiere. Some of the Cinerama films that played at the Eckel included *The Wonderful World of the Brothers Grimm* (1962), starring Laurence Harvey and Claire Bloom, the Stanley Kramer comedy *It's A Mad, Mad, Mad, Mad World*(1964), and, in the Eckel's last shining moments, *2001: A Space Odyssey*, which opened with a special "reserved seats only" engagement on June 26, 1968. Other upstate theaters offering Cinerama included the Teck at 768 Main Street in Buffalo; the Monroe on Monroe Avenue in downtown Rochester; the Panorama in Penfield, a suburb of Rochester; and the Hellman at 365 Washington Avenue in Albany.

Most of the film studio energy that had gone into B pictures for double features now went into television production. Even Columbia Pictures, marginal during the studio era, profited greatly from such TV series as *Rin Tin Tin* and *Circus Boy*, made by its Screen Gems division. One result of the shift to television production was that fewer feature films were being made for theatrical distribution. Foreign films began to fill art house cinemas. These included the British Ealing Studios comedies such as *The Mouse That Roared, Kind Hearts and Coronets* (starring Alec Guinness), and the *"Carry on . . . "* series. The best of the European auteur directors were shown at the Riviera Theater on South Salina Street. These included Ingmar Bergman films from Sweden, such as *Seventh Seal* (1957) and *Wild Strawberries* (1957); Federico Fellini films from Italy, such as *La Strada* (1954), starring Anthony Quinn, and *La Dolce Vita* (1960); and the films of François Truffaut, Jean-Luc Goddard, and the French New Wave.

Drive-In Movie Theaters

Although the number of drive-in theaters has declined in New York, as it has nationally, this increasingly rare cinematic experience is still available in central New York in the twenty-first century. The story of drive-in theaters in New York State begins in 1938 on Long Island, with the opening of the Valley Stream Drive-in, a few miles east of the New York City line at Rosedale, Queens. Closed for World War II, the Valley Stream would not reopen until 1948, when it was renamed the Sunrise Drive-in. Meanwhile, the Kallet Drive-in on Route 5 in Camillus, the first in the Syracuse area, opened in 1946. With the war over and the private automobile back in production—with a vengeance—drive-ins began popping up all over the country during the 1950s. New York reached a peak in the early 1960s, with about 150 drive-ins doing business statewide. The Bay Drive-in in Alexandria Bay and the V Drive-in in Vestal, a Binghamton suburb, both completed in 1968, were among the last "golden-age" drive-ins to be built.

59. The Kallet Drive-in in Camillus, New York. Courtesy of Robert Kallet.

Some of New York State's oldest drive-ins are still in business. The Midway in Minetto (located midway between Fulton and Oswego) was built by Irving and Ruben Canter in 1948 and was later bought and rehabilitated by John Nagelschmidt. Not far behind the Midway in age or charm is the Finger Lakes Drive-in, west of Auburn on Route 5 in Cayuga County, which opened in 1950. The West Rome Twin Drive-in, on West Dominick Street (Route 69) in Rome, opened with a single screen in 1951 and was taken double in 1985 by Conrad and Linda Zurich, who bought it in 1979. The Zurichs also twinned their other drive-in, the Elmira.

The contemporary drive-ins offer patrons opportunities to pick and choose their own balance of nostalgia and convenience. All offer up-to-date sound via low wattage FM frequencies that can be picked up by car radios, boom boxes, or personal stereos. Purists, however, are invited to use a limited number of old-fashioned wired speakers on poles at many of the theaters. The classical drive-in concession stand favorites are available, but so are entire meals. Most have play areas for the kids.

There are no longer any drive-in theaters in Syracuse's immediate suburbs, such as those that surrounded the city in all four directions at one time: DeWitt (east), Camillus (west), Cicero (north), and Nedrow (south). However, with ten outdoor theaters still doing business, May to September, within an evening's drive of Syracuse, the region is relatively well-served, and there is even evidence of a revival. John Nagelschmidt (who had saved the Midway), together with Loren Knapp, bought the Black River Drive-in, on Route 3 between Watertown and Fort Drum, in the mid-nineties. After thoroughly renovating it, they reopened the theater for business in 2006, ending a twenty-year hiatus. Another closed venue, the Bath Drive-in on Route 415 in Steuben County, reopened for the summer of 2007 as Papa's Place Drive-in.

Although legend ties the drive-in to teenage dating, most drive-ins made their money catering to family audiences, offering whole dinners as well as snacks and providing playgrounds for youngsters. Most showed double or triple features, and when it got dark, Mom and Dad could put the kids to sleep in the back seat and enjoy the movie. Drive-ins began their decline in the seventies. The suburbs kept crowding in, sending real estate prices sky-high and making it very difficult for owners to say no to developers. Sumner Redstone, the head of the mammoth entertainment conglomerate Viacom, worked for his father's company, Northeast Entertainment, which owned many drive-ins during the sixties, including the largest in central New York, the Salina, located on U.S. Route 11, just south of Syracuse. Redstone is credited for understanding that the same piece of land that holds a single drive-in movie can be turned into a parking lot with a building in the middle that has ten, twelve, or more screens in it. He is credited with coining the term *multiplex*, which sealed the fate of the drive-in movie.

Downtown Movies No More

Beginning in the fifties and gaining speed in the sixties, the downtown theaters began disappearing, as downtowns in general began to turn into ghost towns after the working day. Since the 1970s, the viewing of films outside of the home is all but limited to the shopping mall multiplexes in the suburbs. The reason can be summed up in a single word: parking. The decline of neighborhood theaters was another blow to urban life. Movie theaters had been the cornerstones of their respective neighborhoods, centers around which community life revolved. With their demise, many neighborhoods lost their identities, a factor in their hastening declines. With three neighborhood theaters in operation, Syracuse has fared better than many American cities. All have persisted as social and commercial cornerstones of their neighborhoods: the Palace in the Eastwood section of Syracuse, the Manlius Art Cinema in the Village of Manlius, and the Hollywood in Mattydale, just north of the city. Seeing a film in these theaters or in the state-of-the-art multiplexes at the Carousel Mall in Syracuse, Shoppingtown in Dewitt, or the Great Northern Mall in Clay, is a different experience from watching a film on television in the isolation of the home. The social dimension of watching in a theater, sitting in an audience of strangers and experiencing emotions in the context of others—laughing at a joke or gasping at a murder—is missing when watching TV at home.

6

❋ The Schine and Kallet Circuits ❋

Exhibition in Central New York

Two theater circuits dominated movie exhibition in central New York State during the mid–twentieth century. The Schine theater chain, with headquarters in Gloversville (Montgomery County), was owned and managed by the Schine brothers— J. Myer Schine and Louis W. Schine. The Kallet theaters, headquartered in the city of Oneida (Madison County), was also owned and managed by two brothers, Myron "Mike" Kallet and Joseph S. Kallet. The Schine circuit was much the larger enterprise, stretching beyond the borders of New York State west into Ohio and Kentucky, south into Pennsylvania, and southeasterly into Delaware and Maryland. At its peak, Schine was the largest independent theater chain in the United States. The Kallet circuit was limited to upstate New York, and concentrated largely in the Mohawk Valley between Utica and Syracuse.

The Schines

In 1902, eleven-year-old Junius (known later in life as "J. Myer") and nine-year-old Louie arrived with their mother from Latvia to join their father, who had settled in Jamestown (Chautauqua County). The family was not wealthy and the boys spoke no English, but at an early age they shared a strong desire to make something of themselves. The brothers worked a succession of jobs in western New York between Jamestown and Buffalo, taking mill jobs and working as candy butchers on the railroad (in those days, candies did not come from the factory pre-packaged, but were cut to order and wrapped at the point of sale—in this case, on the trains). Their goal was to have a business of their own, and their first venture was a newsstand. After Louis enlisted in the army, Myer made a daring leap into the new motion picture business, buying the Novelty Theatre, a Syracuse nickelodeon. The venture collapsed less than three years later. On the train home, returning to Jamestown in apparent defeat, he met a salesman who told him of a

This chapter was written with the assistance of Karen Colizzi Noonan of Geneva, New York, a member of the executive committee of the Theater Historical Society of America.

60. The Glove Theatre in Gloversville, New York, was the headquarters for the Schine operation. It is now the site of the Schine Theatre Museum.

movie house "with possibilities" in Gloversville, a small factory town (where, yes, they made gloves) about a hundred miles east of Syracuse. He could not resist having a look at it; seeing it, he could not resist buying. In 1917, he negotiated a lease for the Hippodrome and proceeded to clean up the place, which was poorly equipped, adding new seats and restrooms. Located on the second floor of the Odd Fellows Temple on East Fulton Street, the Hippodrome opened for business a few months later.

Business was so good that the Schines purchased the relatively new Glove Theatre on North Main Street and took out a lease on the Family Theatre adjacent to it. A decade of expansion for the Schines and the country—the 1920s—followed. The Schines expanded east to Amsterdam, buying the Lyceum (which they renamed the Mohawk), and west to Oneonta, taking over the Strand. By 1929, they had created a circuit of 150 theaters in five states, and while they would buy and sell many properties over the next thirty-five years, they held on to the Glove Theatre, their home office and flagship.

In March 1929, the Schines surprised the movie exhibition industry and the entire upstate business community by selling 98 of their 150 movie houses to Fox Theater Corporation, the exhibition arm of Twentieth Century Fox. Why did the Schines cash out on two-thirds of their theaters at a time when the movie business was booming with the introduction of sound? Whether the result of economic

Main Street, Oriskany Falls, N. Y., Looking West.

61. The Star Theatre, Oriskany Falls, New York. Courtesy of the Limestone Ridge Historical Society.

genius or just plain luck, their timing could not have been better for getting top dollar. The stock market crashed before the end of the year, sending the price of everything south for the next decade.

Flush with capital at a time when it was so scarce, the Schines paid rock-bottom prices for real estate, hotels, radio stations, amusement parks, and retail ventures, building an impressive corporate folio. Schine Enterprises, Inc., purchased miles of prime undeveloped beachfront in Palm County, Florida, including Gold Coast properties in Palm Beach and Boca Raton. They took title to some of the finest hotels in the country, including the Roney Plaza, the most spectacular of the early Miami Beach hotels; the Ambassador Hotel in Los Angeles and its famous Cocoanut Grove nightclub; and, back home in upstate New York, the Gideon Putnam Hotel, an elegant Georgian resort favored by the smart set summering in Saratoga Springs, and the Ten Eyck Hotel on State Street in Albany, where generations of New York politicians gave meaning to the phrase "smoke-filled room."

Despite or because of their differing personalities, the Schine brothers worked well together. Trim and neat, Myer was a cool, analytical, tough-as-nails businessman. He gained a reputation for keen intuition about people, hiring the best and the brightest employees to run the fast-growing Schine empire. Louie, by contrast, was known as a teddy bear, easygoing and charming, plainspoken, a regular guy with a heart of gold. He oversaw the company's day-to-day operations and pressed the flesh with field personnel. From the secretarial pool at the home

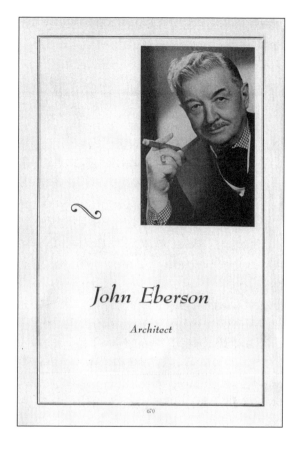

62. John Eberson designed many of the Schine theaters. Photograph from *The 1938 Film Daily Yearbook of Motion Pictures*.

office to the cleaning crew at the farthest flung theater in central Ohio, everybody loved Louie. Either brother without the other would not likely have been nearly so successful.

The Schine chain encompassed a wide spectrum of theater types, ranging from small screens in out-of-the-way places to opulent palaces at urban centers. Entering one Schine theater, patrons might relax in a peaceful Italianate garden, gazing up at stars twinkling among the clouds. In another, crystal chandeliers sparkled overhead, lush tapestries accented the walls, and impressive furniture graced the elegant foyers of the men's and ladies' lounges. Chrome moldings set off stunning deco whirls. Long before there was color on the screen, there was plenty of it in the theater. Bright turquoise, purple, silver, and black were reflected in ornately decorated mirrors. Modern ice trim framed red and black lacquer doors. Lighting fixtures took the shapes of shooting stars and ringed planets. At the entrance, each theater sported the obligatory neon chaser-bulb frenzy of the classic movie marquee. A flea circus or B-picture might be playing inside, but those galloping marquee lights promised the greatest show on earth.

The Schines became important patrons of John Eberson, the grand master of American theater architecture during its grandest period. Eberson designed many of Schines's art deco delights, only a few of which survived the twentieth century. Born in 1875 in Cernauti, Bukovina, a region of current-day Romania, Eberson was already an architect when he immigrated to St. Louis shortly after the turn of the century. Once in America, he began experimenting with the design standards and styles that would become his signature in American history. His first movie theater project was a renovation in Hamilton, Ohio, commissioned by the Jewel Photoplay Company in 1909. Eberson created the Jewel Theatre, which seated 350 in a 40 x 50 foot space inside a mixed-use pre–Civil War building. Soon after, he moved to Chicago, a hub of architectural ideas in the 1910s. With his growing success, he established a base of operations in New York City during the 1920s, living and working there until his death in 1965.

Eberson's greatest theaters are definitive examples of the atmospheric style, a genre of architecture that seeks to transport patrons into another time and place through exotic interior design. Eberson wrote of his technique, "We visualize and dream a magnificent amphitheater, an Italian garden, a Persian court, a Spanish patio or a mystic Egyptian temple yard, all canopied by a soft moonlit sky" (as quoted by Jim Blount, "Jewel Entertained City for 17 Years," *Hamilton Journal–News*, July 12, 2006). The most delightful atmospherics, as Eberson indicates, were representations of the outdoors. He created peaceful garden settings by painting twilight skies with twinkling stars. Grottos with waterfalls and stuffed birds perched on statuary were among the favorite touches of "Papa John," as he was called by colleagues in his later years. Although not the creator of the atmospheric style, Eberson was the architect who brought it to market and made it popular. He was sought out by the Schines as often as possible.

The Schine Auburn Theatre, in the heart of downtown Auburn on South Street, is among Eberson's most exceptional buildings. Closed in 1992, it was promised a new life in 2006 as a performing arts center seats when the Cayuga County Arts Council received a $250,000 federal grant to restore it for that purpose. One can only hope that the restoration of the interior will be loyal to the original, in which the architect melded the atmospheric and art deco styles in a unique outer-space motif. When the theater opened on September 16, 1938, Eberson's unexpected leap from antique Mediterranean luxury to the science-fiction future caused quite a stir.

Foyer and lobby lighting fixtures were shaped like comets and shooting stars. The low, sleek concession stand was accented with heavy infusions of chrome, animating it with an ultramodern metallic aura. Vivid color was abundant, especially in the running bands and stripes on the lobby ceiling, where the requisite Schine red clashed and harmonized with yellow, mauve, chartreuse, and turquoise. In the auditorium, stars and planets floated above in space, and the walls appeared three-dimensional due to the use of the mottled painting technique,

63. The Schine Auburn Theatre, Auburn, New York, designed by architect John Eberson. Photograph from *The 1938 Film Daily Yearbook of Motion Pictures.*

64. Foyer of the Schine Auburn Theatre, showing Eberson's lighting fixtures shaped like comets and shooting stars. Photograph from *The 1938 Film Daily Yearbook of Motion Pictures.*

65. The sleek chromed concession stand in the lobby of the Schine Auburn Theatre. Photograph from *The 1938 Film Daily Yearbook of Motion Pictures.*

which layered, then rag-rolled, midnight blue and deep purple. The ceiling was a shell design, scalloped in tiers to the balcony. Impressive matte chandeliers on either side of the auditorium formed a huge star from which twin comets emerged in a graceful arched trajectory. Their tails were art deco coves in the plaster sky, softly backlit for dramatic effect. The Auburn was the fourth Schine theater in a city that had but four.

The management of movie theaters—the exhibition business—was not for the faint of heart during the studio era. Meyer was a hard-nosed businessman, ready, willing, and able to use the clout of his entire theater circuit to lean heavily on a competitor who might own a single theater. The film distributors demanded minimum cash guarantees and minimum runs for their pictures against a percentage of the overall gross. In addition, exhibitors were expected to contribute to publicity costs. These expenses were variables, subject to frenzied negotiations; there were no standard contracts. A smart deal on a picture could mean the difference between profit and loss. Finally, there was the "house nut"—the monthly expenses of running a movie theater: payroll, utilities, equipment, insurance, building maintenance, and so on. When two or more competitors wanted a picture they thought could be a hit, things could get nasty. And of course, there was no guarantee of what any picture might gross. Not even John Wayne could prevent a weekend snowstorm.

66. Celestial decorations in the auditorium of the Schine Auburn Theatre.
Photograph from *The 1938 Film Daily Yearbook of Motion Pictures.*

In May 1942, some thirteen years after the fortuitous sale of most of its the-aters to Fox, the Schine circuit (made up now by a tangle of subsidiaries designed to maximize tax advantages and minimize visibility) had been replenished from 52 theaters in 1929 to just about its former size: 148 theaters across six states, more than half of them in New York. But the statistic that had the most salient impact on the company's future was this: Of the seventy-six towns where Schine operated, sixty were "closed" towns where Schine, with one or more theaters, was the only exhibitor of major motion pictures. This led a federal court to find a conspiracy to restrain trade between the Schine firms and the eight leading distributors of Hollywood films. The court held that through use of its bargaining power, Schine was able "to dictate terms . . . [and] . . . to exert pressure on the dis-tributors to obtain preference." U.S. Supreme Court Justice William O. Douglas, appointed by Roosevelt in 1939, summarized the district court findings as follows: "Through the use of such buying power Schine arbitrarily deprived competitors of first[-] and second-run pictures, [and] was able in many towns to secure unrea-sonable clearances [rights to exclusive exhibition] of from 90 to 180 days." (This quotation, together with other information on the case, appears in *Antitrust in the Motion Picture Industry* by Michael Conant [Berkeley: Univ. of California Press, 1960]). Other questionable practices alleged in the case include the unrea-sonable cutting of admission prices to drive small operators out of business and

67. The Strand was the successor to the Carroll Theatre at 114 E. Dominick in Rome, New York. Photograph by C. B. Howland for the Strand, 1925. Courtesy of Robert Kallet and Joe Pfeiffer.

68. The Star Theatre was an early theater in Rome, New York. Courtesy of Robert Kallet and Joe Pfeiffer.

the buying out of competitors who agreed to sign contracts forbidding them from entering towns designated by the Schines.

In 1942, the federal district court entered a decree enjoining these practices and requiring the Schine brothers to divest fifty theaters located in forty towns in order to turn "closed towns" into towns with competing theaters. Through effective negotiation with the court, J. Myer Schine was able to get the court to reduce the divestiture from fifty to thirty-nine theaters. The case should have ended there—but it did not. In March 1954, the federal government opened civil and criminal contempt proceedings against the Schine subsidiaries and five individuals, including the Schine brothers, charging the defendants with failing to dispose of twenty-three of the thirty-nine theaters that they had been ordered to sell and with conspiring to retain indirect control of some of the sixteen theaters they purportedly had sold. In December 1956, more than seventeen years after the case was first brought, the defendants were held in criminal contempt for having violated the divestiture decree. This decision was affirmed and finalized in 1959, but no penalties were ever meted out. The Schines sold off their entire movie exhibition enterprise, valued at $150 million, to Realty Equities Corporation of New York City in September 1965, retaining only one property: the Schine building in Gloversville, which contained company offices and the Glove Theatre.

The Kallets

In 1910, M. J. "Mike" Kallet and J. S. Kallet, both native Syracusans, rented a store in the Onondaga Valley to operate as a nickelodeon. Dreamland, as they called it, was located on Seneca Turnpike at the end of the trolley line from downtown Syracuse, just down the road from the Valley Theatre, an outdoor venue under canvas, where operas were staged in the summer and movies were sometimes screened.

On a hot July 4, 1910, M. J. was out in front of the Dreamland, a barker with a megaphone cajoling the public, while J. S. was inside, hand-cranking the projector, which was already obsolete by industry standards. Despite the ballyhoo, receipts at the end of the day totaled no more than eighty cents. The Kallets kept Dreamland open for less than a year, moving on to manage theaters all over central New York: Camden, Hamilton, Canastota, Baldwinsville, Weedsport, and Port Byron. Despite their share of hard times, they ended up working at the Grand Opera House in Syracuse, part of the B. F. Keith vaudeville circuit. When M. J. became manager at the Grand Opera House, J. S. moved to the Empire Theatre on South Salina Street as treasurer. He had considerable work to do in 1916 when the Empire offered the Syracuse premiere of D. W. Griffith's *Birth of a Nation* for a two-week run at the unheard-of admission price of two dollars per ticket.

Capitol Theatre   Rome, N. Y.

69. The Capitol Theatre, at 218 West Dominick Street in Rome, New York, was the premiere Kallet theater.

The brothers finally found their way back into a business of their own in 1922, building the Madison Theatre in Oneida, the first in what would become the Kallet chain. J. S. also worked as manager of the Family Theatre in Rome, and within a few years the brothers purchased it, along with two other Rome theaters: the Carroll (which they renamed the Strand) and the Star. They officially incorporated as Kallet Theatres, Inc., in 1925, with offices in the Madison Theatre. In 1928, the Kallets built the Capitol Theatre on West Dominick Street in Rome, their fourth screen in the city. During the 1950s, they operated another family enterprise, radio station WKAL, in studios above the theater in the Capitol building.

The circuit continued to grow, reaching a peak of fifty-seven theaters in the early fifties, most in central New York, although some, such as the Strand in Brockport (see plate 13), the Palace and Riviera in Geneseo, and Farman's and the O-At-Ka in Warsaw, were further afield. The Kallet Theatre on Main Street in Oneida (see plate 14), the second Kallet screen in the city, opened in 1938. With 1,220 seats, it was among the Kallets' most ambitious projects and served as Oneida's premiere film venue until it died a slow death—along with much of downtown Oneida—at the hands of suburban competitors. After several years as a roller-skating rink, followed by vacancy, the theater was reopened as the Kallet Civic Center in 1983, and was the site of live music and dance events. A visit to its Web site

70. Farman's Theatre, part of the Kallet chain, was located on Main Street in Warsaw, New York. It is currently being operated as the Cinema Theatre. Courtesy of the Warsaw (N.Y.) Historical Society.

in 2007, however, showed a calendar with no events listed after 2005. The 558-seat Kallet Theatre on North Jefferson Street in Pulaski (Oswego County) was designed by noted architect Milo Folley and built on the site of the former Temple Theatre, which burned down in 1934. The Pulaski Kallet started showing movies in 1935 and stopped in 1982. However, Pulaski Kallet endures as what might be the world's only art deco auto parts store.

The Regent Theatre, on East Genesee Street in Syracuse, near Syracuse University, was for years the main Kallet screen in the city. As a "second-run exclusive house," it received films nine to twelve days after their downtown runs and two weeks before neighborhood theaters. In 1958, the Kallets sold the Regent to the university, which renovated it to become the drama department and the site for Syracuse Stage, the area's leading theater for live professional stage productions. Other Syracuse theaters operated by the Kallets at various junctures include the Avon on Hawley Avenue, in what is now known as the Hawley-Green neighborhood; the Harvard on Westcott Street, later renamed the Westcott Cinema; and the Riviera, an Italianate atmospheric building at 3116 South Salina Street on the South Side.

In 1946, the Kallets opened the Kallet Drive-in, their first drive-in theater and the first in metropolitan Syracuse, in Camillus, a western suburb. The eight-

71. The O-At-Ka Theatre was another Kallet theater on Main Street in Warsaw, New York. It now houses a Rex-all Pharmacy. Courtesy of the Warsaw (N.Y.) Historical Society.

72. Franklin Theatre on South Avenue in Syracuse. Photograph by George Read.

73. The Regent Theatre, part of the Kallet chain, was sold to Syracuse University in 1957. It is now the site of Syracuse Stage.

hundred-car facility, located at Kasson Road and West Genesee Street, had a striking waterfall cascading over the back of the screen. It did good business but lasted only until 1960, when suburban sprawl made the property too valuable to keep from the hands of developers. A drive-in theater on Middle Settlement Road in New Hartford (Oneida County) was built by a broadcaster in 1950 as the WGAT Drive-in, serving as an advertisement for his radio station as well as an outdoor movie theater. The Kallets bought it from WGAT in 1955, renaming it Kallet's Drive-in, which gave it an identity somewhat distinct from the nearby Kallet Drive-in in Utica on the Utica-Rome Road.

The Kallets can also lay claim to the first movie theater inside a mall in the Syracuse area, which opened for business as the Shoppingtown Theatre in the basement of Shoppingtown Mall in 1955. It was redeveloped by subsequent owners into a multiplex, the Shoppingtown 10. In 1968, the Kallets, who had been pushed out of the mall, constructed a free-standing building adjacent to the Shoppingtown parking lot. It contained two 800-seat theaters equipped with 70 mm projectors, probably in the hope it would become a Cinerama venue. By the early eighties, that hope had been dashed by the lack of Cinerama product, and the theater was subdivided into four 350-seat auditoriums. In 1996, the four-screen Shoppingtown, the last theater built by the Kallets, met the same fate that the

74. The Shoppingtown Theatre, built in the Shoppingtown Mall on Erie Boulevard in Syracuse, New York, was the first of the mall theaters in the region. Photograph by George Read.

75. Another Kallet theater, the Genesee in Syracuse, New York, was torn down in 2001. Photograph by George Read.

brothers' first theater, the Madison, had met in 1958: It was demolished. Although dwarfed by the mighty Schine circuit, the Kallet brothers outlasted the Schine brothers as movie exhibitors, holding on, if just barely, into the twenty-first century. The Kallets finally left the business in 2001, when they sold the Genesee, which was located in Westvale Plaza, an aging outdoor shopping center in the town of Geddes, just past the Syracuse city line. The Genesee was demolished for an auto parts store that lasted only one year.

7

Neighborhood Theaters in the City of Syracuse

During the twentieth century, more than one hundred neighborhood theaters opened and closed their doors within the Syracuse city limits. Before television kept people at home, and before automobile travel made easy parking as important a consideration as what was playing on the screen, every neighborhood, it seemed, had at least one theater. Many of the earliest theaters, the nickelodeons that opened before World War I, were simple storefronts that lasted only a few years. Others endured, and with the soda shops, restaurants, and newsstands that coalesced around them, they became the social cores of neighborhoods, places where neighbors met and friendships were forged and maintained. In Syracuse, as elsewhere in America, theaters were part of the social glue that held urban neighborhoods together. When they disappeared, there was nothing to replace them.

The North Side

In the early days, a string of nickelodeons sprang up along North Salina Street. They included Morgan's (later renamed the Majestic) at 246 North Salina; the Rex at 464; the Star at 547, Luna (later the Princess) at 626; the Studio at 839, and the Crystal at 1636. Other north side nickelodeons could be found on Park Street: the Pictorial (later the Hyperion and later still, the Highland) at 327 Park Street and the Park (later the Rialto) at 1003 Park Street.

Turn Hall, at 621 North Salina Street between Ash and Division streets, was located at the hub of the north side's large German and Italian immigrant neighborhoods, which were known for their fine ethnic restaurants and beer gardens. This stretch of North Salina Street still contains a variety of Italian eateries and espresso bars. Run by the German American Turn Verein, a social and cultural organization, Turn Hall offered German and Italian films during both the silent and sound eras, and was probably the only venue in the city's history that regularly offered a schedule of foreign films. The building burned down in 1954 and was replaced by a social hall and gymnasium.

A bit further to the north was the Globe Theatre at 710 Kirkpatrick Street, between Spring and First North streets. It was the first theater acquired by S. P. "Sam" Slotnick, who eventually assembled a chain of 148 theaters, mostly outside New York State. The building remains as the office of a home remodeling business. Slotnick properties in Syracuse included the Acme Theatre at Butternut and Park streets; the Schiller Park Theater on Grant Boulevard, next to Steigerwald's Meat Market; the East Theater (previously known as the Roxie) on Manlius Street in East Syracuse; and the Groveland Theater on Mill Street in the Limestone Plaza section of Fayetteville. Slotnick drive-ins included the DeWitt on Erie Boulevard, the Lakeshore in Liverpool, and the North in Cicero. In 1983, Herbert Slotnick, the founder's son, sold off the theater chain (by then known as CinemaNational) to concentrate on another family business, fast-food restaurants.

The Lyric, another north side venue, was an upstairs theater in a building on the corner of Wolf and Park streets. Patrons entered from the front (i.e., from behind the screen). The Avon Theatre, at 443 Hawley Avenue, between North Crouse Avenue and Howard Street, was redeveloped into a block of buildings holding retail businesses at street level and apartments above. Others, long gone, include the Lincoln, on Hawley across Lodi Street, and the Melva (later renamed the James), on James Street in Eastwood.

Of the many theaters that once stood north of downtown Syracuse, there are but two survivors still doing business: the Hollywood Theater in Mattydale on U.S. Route 11, which shows second-run films, and the Palace Theatre, at 2384 James Street in Eastwood. The Palace is a revival house, offering many difficult-to-see films as well as live events, including musical acts. In the spirit of neighborhood theaters, the revitalized Palace Theatre and Café features a coffee bar and restaurant where moviegoers and others can mingle.

The West Side

There were two theaters on South Geddes Street, a principal north-south thoroughfare of the west side: the Cameo, near Grand Avenue, and the Langan, which was also known as Tony's Opera House. Tipperary Hill, the city's principal Irish neighborhood, was served by the Burnet Park Theater, near the present-day site of the Rosamond Gifford Zoo. The Liberty Theatre, located on Park Avenue at Liberty Street, across the street from the Frazer School, was in the heart of the west side's Polish enclave, and occasionally screened Polish-language films. James Kernan, who owned the Lyric on the north side, also owned the Kernan on Willis Avenue near West Genesee Street, as well as the Capitol and the Starland. The Kallet family was active in the area after World War II, building one of the first drive-ins in the country in Camillus in 1946 and the Genesee Theater in Westvale

76. Syracuse's Globe Theatre was the headquarters for S. P. Slotnick. Courtesy of Joseph Detor.

in 1950. The Village of Solvay, just across the city line from the west side, was built around the Allied Chemical plant. Its ethnic population was divided into groups: Tyroleans from the north of Italy and Sicilians from the island province in the south. They were served by two Solvay theaters, the Allen on First Street and the Community on Milton Avenue.

The East Side

The east side includes Syracuse University and a student neighborhood stretching east from campus along Euclid Avenue and north along Westcott Street. A large number of professionals live in the area as well. The area's exhibition history begins in the nickelodeon period with the opening in 1913 of the Crouse at 112 Crouse Avenue and the Irving at 823 E. Genesee Street, followed by the Varsity at 327 Irving Avenue and the Little at 704 E. Fayette Street. The city's Jewish immigrant quarter, located between Syracuse University and downtown Syracuse until the building of Interstate 81 displaced many of the residents, was served by the Alcazar Theatre on present-day Oakwood Avenue (then known as Grape Street) near Raynor Avenue. Abe Korren, affectionately known as "Pop Corn," ran that theater for many years. The Regent Theatre on East Genesee Street was saved from the wrecking ball by Syracuse University, which converted

77. The Cameo Theatre stood at 656 South Geddes Street in Syracuse.
Courtesy of Richard G. Case, *The Herald Journal*, February 17, 1967.

it to Syracuse Stage. The Westcott Cinema on Westcott Street, in the heart of
the student area, remained an independent theater screening rarely shown indies,
foreign films, and documentaries until October 2007, when rising costs forced its
closure. It was originally known as the Harvard Theatre, where Syracuse University's drama department had its Boar's Head Theatre company productions. In
the suburbs to the east, Fayetteville had the Slotnick family's Groveland Theater
on Mill Street and later a multiplex at the Fayetteville Mall. Both are gone, leaving Fayetteville without a cinema. In the neighboring Village of Manlius, the
Manlius Art Cinema at 135 East Seneca Street is going strong. Under the same
management as the now-defunct Westcott, it offers similar fare. The Manlius was
once known as the Strand.

The South Side

Nickelodeons were located all along South Salina Street, the route of the downtown streetcar. They included the Mecca (at 1010), the Lafayette (at 2622), and
the Arcadia (at 2203) near West Colvin at South Salina. As was the case with the
Lyric, patrons entered the Arcadia Theatre from the front. The Brighton (at 2610),
had 1,750 seats, an enormous capacity for a neighborhood house. Still further

78. The Community Theatre, shown here c. 1948, was on 1725 Milton St. in the Village of Solvay. Courtesy of Lorenzo T. Fernandez.

south (at 3116) was Harry Gilbert's Riviera, an "atmospheric" theater, done in a beautiful Moorish decor and sporting a star-studded ceiling. The Riviera had been surviving as an art house until 1968, when its roof caved in from heavy snows. Two more theaters were located on South Avenue, another of the area's main arteries: the Franklin, at 237, and the Elmwood on the 400 block.

Downtown Syracuse

Until suburbanization robbed so many American cities of their nightlife, the downtown area of any city usually had the most movie theaters. Syracuse was no exception. Downtown, South Salina Street was lit up like a virtual Broadway, with Loew's State, RKO Keith's, the Paramount, and the Empire, with the Strand nearby on the corner Salina and Harrison streets, across from the Hotel Syracuse. The Crescent, the earliest downtown theater, was built on South Salina in 1909 on the site now occupied by the Galleries shopping mall. Fayette Street, the main east-west artery, had the Eckel on East Fayette, and the Novelty and the Rivoli opposite each other on West Fayette. On Warren Street stood the Standard in the Bastable Building (now the State Tower Building) and the Savoy (next to the *Syracuse Journal* building), which later became a burlesque house. The Happy Hour (later the Swan, and still later the Mid-Town) was on the first block of North Salina Street, across from the present Syracuse *Post Standard* building. The Seymour was in an old church at 159 Seymour Street. The Civic (later known as the Syracuse, the Top, the Ritz, and the System) was at 572 South Salina.

Between 1910 and 1920, before the movie palaces were built, nickelodeons could be found downtown on South Salina: May's (328), the Palace (429), the Majestic (473), the Antique (476), and the Salina (544). Other nickelodeons included the Dreamland at 110–120 W. Onondaga St., the Colonial at 1234 S. State Street, the Crystal at 511 Eagle Street, and the Eagle at 942 Eagle Street.

As noted earlier, movie house attendance declined steeply after hitting its peak in 1947. This was not because of television, which was still five years in the future, but because a monopoly case brought before the U.S. Supreme Court in 1947 ended the studio system of movie exhibition. The case, *U.S. v. Paramount*, found against the major Hollywood studios and decreed that they must divest themselves of one of their three arms of business—production, distribution, and exhibition. The studios, taking the easy route, sold their theater chains. Now lacking a ready market, the studios cut back on production—first the "B" film units, then short films, newsreels, and cartoons, and finally, serials, all the staple diet of the neighborhood theaters.

By the early 1950s, unable to survive on reruns alone, neighborhood theaters started shuttering right and left. The downtown theaters followed ten years later. The theaters were what brought people downtown at night. The department stores could not do this on their own, especially with competition from the newly emerging suburban malls, where parking was convenient. So downtown Syracuse, at night, became a ghost town.

Thus, an era came to an end. And all we have left are our memories.

1. Weiting Opera House.

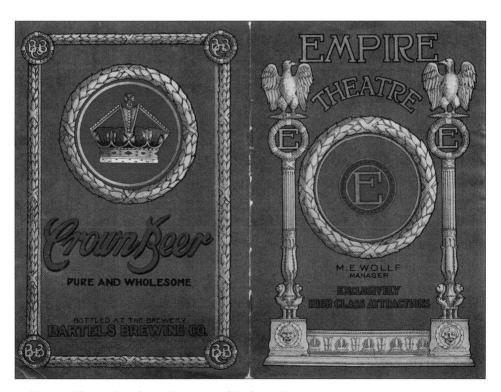

2. Empire Theatre brochure. Courtesy of Barbara Opar.

3. Turn Hall. Courtesy of the Syracuse Turners.

4. Ladies' Room, B. F. Keith's Theatre. Courtesy of David Jenks.

5. Foyer, B. F. Keith's Theatre. Courtesy of David Jenks.

6. Boxes and balcony, B. F. Keith's Theatre. Courtesy of David Jenks.

7. Westcott Cinema today.

8. Brighton Theatre today.

9. Grand Staircase, Loew's Theatre. Courtesy of Syracuse Area Landmark Theatre, Inc.

10. Auditorium from stage, Loew's Theatre. Courtesy of Syracuse Area Landmark Theatre, Inc.

11. Grand Promenade today, Loew's Theatre. Courtesy of Syracuse Area Landmark Theatre, Inc.

12. Fox Movietone truck. Collection of the Case Research Lab Museum.

13. Strand Theatre, Brockport. Photo by Joseph Pfeiffer, Jr.; courtesy of Robert Kallet.

14. Kallet Theatre, Oneida. Photo by Joseph Pfeiffer, Jr.; courtesy of Robert Kallet.

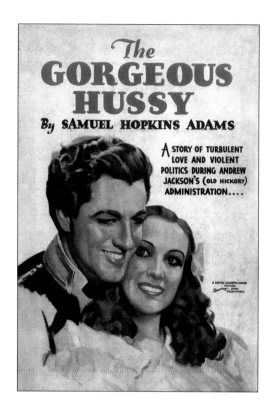

15. Book cover of *The Gorgeous Hussy.* Courtesy of the Penguin Group (USA), Inc.

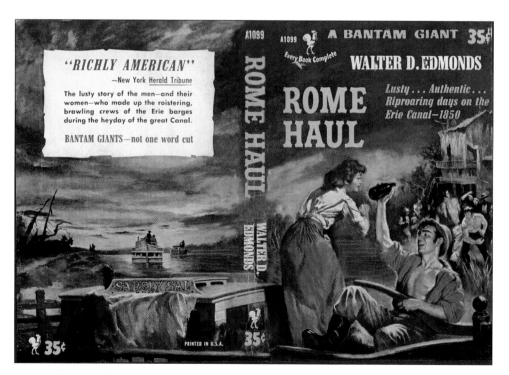

16. Book cover of *Rome Haul.* Courtesy of Random House, Inc.

17. Advertisement for film of *Drums Along the Mohawk*. Copyright © 1939 Twentieth-Century Fox. All rights reserved.

18. Book cover of *Chad Hanna*. Courtesy of Little, Brown and Company, a division of Hachette Book Group USA, Inc.

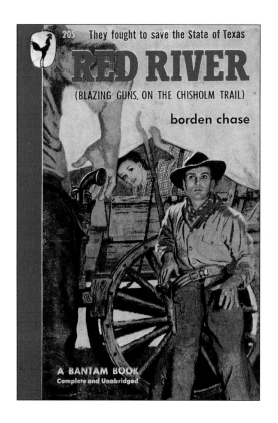

19. Book cover of *Red River*. Courtesy of Random House, Inc.

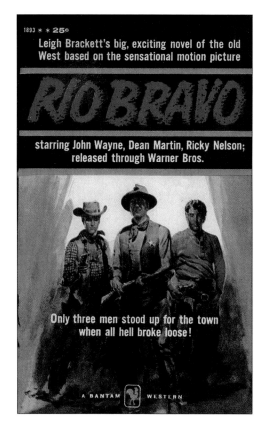

20. Book cover of *Rio Bravo*. Courtesy of Random House, Inc.

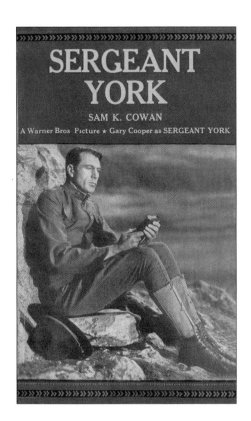

21. Book cover of *Sergeant York*. Courtesy of Penguin Group (USA), Inc.

22. Book cover of *David Harum*. Courtesy of Penguin Group (USA), Inc.

23. Book cover of *Tess of the Storm Country.* Courtesy of Penguin Group (USA), Inc.

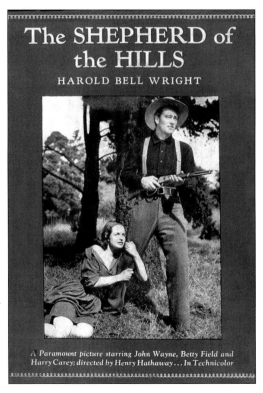

24. Book cover of *The Shepherd of the Hills.*
Courtesy of Penguin Group (USA), Inc.

PART TWO

People

8

Cinema Figures with Links to Central and Upstate New York

A Biographical Dictionary

Central New York State has played a significant role in the technological, industrial, and artistic development of American cinema by spawning, nurturing, and inspiring an impressive number of influential writers, producers, directors, performers, and businesspeople. Some are native; others came to teach or to be educated at the area's renowned institutions. Many came upstate during the formative era of the movie industry, making it an early center of experimentation and production. What follows is an eclectic—and no doubt incomplete—listing of personalities, ranging from popular stars who happened to be born in central New York, to writers and artists who endowed the cinema with the cultural legacies of the region.

GEORGE ABBOTT (1887–1995), playwright, screenwriter, director, producer, was born in Forestville (Chautauqua County). A graduate of the University of Rochester, he began his career on Broadway, where he wrote, directed, and produced a considerable body of works. Turning to screenwriting at the dawn of the talkie era, he is credited with more than forty features, adapting many of his own plays for the movies, including *Broadway* (1929; remade 1942), *Three Men on a Horse* (1936), and *The Boys from Syracuse* (1940). Nominated for three Academy Awards, he took home an Oscar in 1930 as a collaborating writer on *All Quiet on the Western Front*. Abbott wrote and directed the stage and screen versions of two of Broadway's biggest hit musicals of the postwar era, *Damn Yankees* (1957) and *The Pajama Game* (1958). He lived to the age of 107.

HELEN ABERSON (1907–1999), writer of children's fiction, was a Syracuse native and a 1929 graduate of Syracuse University. She sold her short story, "Dumbo, the Flying Elephant," to the Disney studio for one thousand dollars, and it became the source for *Dumbo* (1941), an animated feature about a flying elephant that quickly became a children's classic. Some film historians claim that *Dumbo*'s success came at a crucial moment, saving Walt Disney Pictures from financial ruin. Harold Pearl, Aberson's husband and collaborator, was manager of the Eckel Theatre on East Fayette Street in downtown Syracuse.

SAMUEL HOPKINS ADAMS (1871–1958), journalist and novelist, was best known for his investigative journalism, especially a series of exposés on the patent medicine industry that helped spur congressional passage of the 1906 Pure Food and Drug Act. But he was also a prolific short fiction writer, whose stories were the sources for sixteen films. The most famous of these is *It Happened One Night* (1934), a romantic comedy based on Adams's "Night Bus," a story about a wealthy young heiress who flees her constricting family and falls in love with a reporter who at first is more interested in her life story than her. The film dominated the 1934 Academy Awards, winning Best Picture, and gaining Oscars for director Frank Capra and stars Claudette Colbert and Clark Gable. "Night Bus" also inspired a Hollywood musical, *You Can't Run Away From It* (1956), directed by Dick Powell and starring June Allyson and Jack Lemmon, as well as at two international productions: *Nina Bonita* (Philippines [Tagalog language], 1955) and *Aslan Marka Nihat* (Turkey, 1964). Born in Dunkirk, near Buffalo, Adams was an 1891 graduate of Hamilton College in Clinton, where he organized the school's first varsity football team. He stayed in the region, making his home for many years on Owasco Lake in the eastern Finger Lakes, south of Auburn. Writing under the pseudonym Warner Fabian in the 1920s, Adams explored the sexual awakenings of young women during the jazz era in two novels that were brought to the screen: *Flaming Youth* (1923)—also the title of the film, starring Colleen Moore—and *Unforbidden Fruit* (1929), which was the basis for *The Wild Party* (1929). The latter was directed by Dorothy Arzner and starred Clara Bow (the "It" Girl) as a college coed whose major interests are anything but academic. The film outraged would-be censors and is often cited as a factor in the lead up to the creation of the Hollywood "code." Adams's stories served as sources for a succession of hits featuring major stars, including *The Gorgeous Hussy* (1936; see plate 15) with Joan Crawford and Robert Taylor; *The Perfect Specimen* (1937), which gave Errol Flynn his first leading role; and *The Harvey Girls* (1945) starring Judy Garland.

HERVEY ALLEN (1889–1949), fiction writer, was a Pittsburgh native who lived in Syracuse and, later, in nearby Cazenovia. Allen wrote *Anthony Adverse*, a novel set in eighteenth-century Italy concerning the life of an unwanted son who is sent as an infant to be brought up at a convent. The 1936 film, starring Fredric March and Olivia de Havilland, won four Academy Awards and was nominated for three more.

HAROLD ARLEN (1905–1986), songwriter, was born Hyman Arluck, the son of a Buffalo cantor who, like Al Jolson in *The Jazz Singer,* chose a secular musical career. As a young musician, Arlen led a dance band, the Buffalodians, which played in nightclubs and on Lake Erie steamers. Making a name for himself on Broadway, he worked with the elite of Hollywood lyricists. "Get Happy," written with Ted Koehler for *The Life of the Party* (1930), was on the soundtrack of six MGM musicals in the 1930s. Although not credited on screen, Arlen collaborated with Yip Harburg on songs for *The Wizard of Oz* (1939), and the pair won the Academy Award for "Over the Rainbow." Arlen's tunes are among the most familiar in the American songbook, including such standards as "Stormy Weather," "I've Got the World on a String," "Come Rain or Shine," "That Old Black Magic," and "One for My Baby." His name appears in the credits of more than 250 films and television programs.

IRVING ADDISON BACHELLER (1859–1950), journalist, fiction writer, and screenwriter, was a native of Pierrepont (St. Lawrence County) and graduated in 1882 from nearby St. Lawrence University in Canton. Moving downstate to seek his fortune, he founded the first American newspaper feature syndicate in 1884, serializing Joseph Conrad, Arthur Conan Doyle, Rudyard Kipling, and other British authors for American newspaper readers. By 1898, he was an editor at the *New York World.* But his real passion was writing fiction and poetry. Bacheller's *Eben Holden: A Tale of the North Country* (1900) sold more than a million copies at the turn of the century and is valued by historians as an authentically detailed picture of Adirondack life in the nineteenth century. Bacheller wrote the screenplay for *The Light in the Clearing* (1921), adapting it from his 1917 novel of the same title. A pre–Civil War story, it was set in the north country during the political reign of Silas Wright, a Canton lawyer who rose to become governor. Two of Bacheller's poems became title subjects for MGM shorts: "Whisperin' Bill," about a congressman who visits a farmer to ask for his vote; and "Shep," concerning a man forced to put down his beloved aging dog.

KING BAGGOTT (1874–1948), actor, was a towering leading man who starred in silent adventure dramas. In *The Eagle's Eye* (1918), a patriotic Wharton Brothers film financed by William Randolph Hearst and made in Ithaca, he singlehandedly breaks up a nest of German spies. Baggott was also active on the World War I home front: He sold Liberty Bonds at local rallies, and he appeared in *Mission of the War Chest* (1918), a silent agitprop documentary made by the Tompkins County War Chest to raise money for the boys "over there."

79. King Baggott, leading man in many silent films, sits between stage actress Josephine Victor, left, and screen actress Maidel Turner, c. 1900. Library of Congress, Prints & Photographs Division, LC-DIG-ggbain-18035.

ALEC BALDWIN (1958–), DANIEL BALDWIN (1960–), WILLIAM BALDWIN (1963–), and STEPHEN BALDWIN (1966–), actors and brothers, were all born and raised in Massapequa (Nassau County) on Long Island. None attended college in upstate New York, and the closest any of them lives to the region is Nyack (Rockland County) in New York City's northern suburbs. Why then are the Baldwins, who have collectively starred in more than one hundred feature films, often photographed by paparazzi in the Syracuse area? The answer is simple: They are good boys and visit their mother, Carol Martineau Baldwin, a Syracuse native and graduate of Syracuse University, who moved back to the area after the death of her husband Alexander in the 1980s. A survivor of breast cancer, she has worked ever since to fight for a cure through the central New York chapter of the Susan G. Komen Breast Cancer Foundation in Syracuse and the Carol M. Baldwin Breast Cancer Research Fund, which supports cancer research at the SUNY Stony Brook Medical Center on Long Island.

LUCILLE BALL (1911–1989), actress, was born in Jamestown (Chautauqua County). Following her father's death from typhoid in 1915, she was raised at her maternal family's farm on the outskirts of town. From early childhood, Ball expressed desires to become an entertainer and was encouraged by her mother and grandparents. She left high school at age fifteen to attend dramatics school in Manhattan but came home after six weeks, intimidated by the other students, who included Bette Davis. It was the first of several false starts on Broadway for Ball. While working as a model in New York City, she was spotted by Eddie

Cantor, who was about to make his first picture, *Roman Scandals* (1933), and accepted his offer of an uncredited chorus part as a Goldwyn Girl. She arrived in Hollywood in 1933 as an MGM contract player, appearing in more than seventy features, most of them B comedies. She often played a chorus girl, as in *Kid Millions* (1934) and *Ziegfeld Follies* (1946), but on occasion won roles as a wisecracking second banana for big-name comedians such as the Marx Brothers in *Room Service* (1938) and Bob Hope in *Fancy Pants* (1950). In the television era, Ball achieved worldwide fame as the star of *I Love Lucy*, the most popular situation comedy of the 1950s. Throughout her career, she maintained strong ties to her hometown. Her characters on *I Love Lucy* and her later television series were all written as Jamestown natives, and she was a loyal contributor to several local charities, establishing a scholarship fund in her name at Jamestown Community College. In 1956, Ball held the world premiere of *Forever Darling*, a feature film she made with Desi Arnaz, at the Palace Theatre in Jamestown. The city is home to a museum dedicated to Ball's life and career and holds an annual "Lucy-Desi Festival" on Memorial Day weekend. Ball is buried at Lakeside Cemetery in Jamestown.

PHILIP BARRY (1896–1949), playwright, hailed from Rochester. His father, a wealthy importer of marble and tile, died a year after his birth, and he was raised by his mother, a member of a "lace-curtain Irish" Philadelphia family. Barry's 1939 Broadway hit, *The Philadelphia Story*, was set in the upper crust of Main Line society that his mother had described to him in detail. Adapted for the screen by Donald Ogden Stewart and assigned to director George Cukor, *The Philadelphia Story* (1940) is widely considered among Hollywood's greatest romantic comedies. Katharine Hepburn and Jimmy Stewart both won Best Actor Oscars, and Cary Grant shines in a supporting role. The American comedy of manners was a Barry specialty. Cukor had directed Grant and Hepburn in a Stewart screenplay of Barry's *Holiday* in 1938. Other Barry plays that came to the screen included *The Animal Kingdom* (1932), starring Leslie Howard and Myrna Loy; *Without Love*, starring Hepburn and Spencer Tracy; and *One More Tomorrow* (1946), starring Ann Sheridan. After Barry's death, many of his plays were produced for television during the medium's golden age of anthology drama.

LIONEL BARRYMORE (1878–1954), actor, was the brother of Ethel and John Barrymore (with whom he appeared in *Rasputin and the Empress* [1932]), and great-uncle to contemporary star Drew Barrymore. He began his screen career in Ithaca as the villain in *The Exploits of Elaine* (1914), a Wharton Brothers serial. Bookish and aloof in his personal life, he was nicknamed "The Professor" by members of the Ithaca film colony who thought he looked more at home "high above Cayuga's waters" on the Cornell campus than down at the Wharton Brothers's lakeside film studio. During the sound era, Barrymore became a fixture in MGM's galaxy of stars, often playing sentimental grandpas or crotchety millionaires. He is best remembered as the irascible Dr. Gillespie, mentor to Dr. Kildare in fourteen films made between 1939 and 1947, and as the greedy banker, Mr. Potter, the nemesis of Jimmy Stewart in *It's a Wonderful Life* (1946).

LYMAN FRANK BAUM (1856–1919), writer of children's literature, was born in Chittenango (Madison County) and raised in nearby North Syracuse. He married Maude Gage of Fayetteville, the daughter of suffragette Matilda Joslyn Gage, a close associate of Elizabeth Cady Stanton, Susan B. Anthony, and other women's rights pioneers. On a bumpy career path that led him through turns as a playwright, shopkeeper, newspaper reporter, and traveling salesman, Baum seemed to find his calling in midlife as a children's writer, successfully collaborating with several illustrators, including William Wallace Denslow, with whom he completed *The Wonderful World of Oz* in 1900, the best-selling children's book in the world for more than two years. Baum continued to write Oz stories for the rest of his life. Sensing the suitability of his vision for the cinema, he moved to Hollywood in 1910 and founded the Oz Film Manufacturing Company in 1914, producing nine of the fourteen Oz stories for the screen. However, he had not understood that while children's literature was a genre with a well-developed market constituency, no children's format had yet developed in the nascent film industry, and he sold the studio to its next-door neighbor, Universal Pictures, in 1915. Baum died several years later, not knowing that his name would appear in the credits of more than 125 films in a dozen languages during the next century. His best-known children's story, "The Wonderful Wizard of Oz," reached the screen as *The Wizard of Oz*, first for the silent screen (1925), and later in the sound era as a Hollywood musical starring Judy Garland, the version that would become one of the greatest blockbusters ever produced. The spectacular use of Technicolor to express the fantastic nature of Oz was one of the first commercial uses of the technique. Although the story putatively begins in Kansas, the sepia segment that opens the film contains shots of a gathering storm that looks very much like the Great Lakes weather systems that come rolling across the central New York skies of Baum's childhood. Chittenango celebrates Baum's birthday each summer with Ozfest, a weekend-long celebration that includes a parade held on Main Street, which is painted as a yellow brick road. Surviving members of the cast, including many of the performers who played the Munchkins, attend the event. The Baum story also served as the source for *The Wiz* (1978), starring Diana Ross and Michael Jackson, an African American retelling of the story in a Harlem setting.

ANDREW BERGMAN (1945–), screenwriter, director, and producer, was born in Queens and attended Binghamton University, graduating *magna cum laude* in 1965. A comic writer who specializes in satirizing popular genres, he wrote the story and collaborated with Mel Brooks on the screenplay for *Blazing Saddles* (1974), and wrote scripts for such comedies as *Oh, God! You Devil* starring George Burns (1984) and *Soapdish* (1991). Bergman directed his own script for *The Freshman* (1990), a *Godfather* spoof starring Marlon Brando and Matthew Broderick.

LOUISE BROOKS (1906–1985), actress, is sometimes called "the lost star of the twenties." She dazzled as a bright comet in the Hollywood sky at Paramount in the late 1920s, notably in *Beggars of Life* (1928) with Richard Arlen and Wallace Beery, and *The Canary Murder Case* (1929) with William Powell. Her black bobbed hair and bangs and penetrating eyes created a mesmerizing image of the new American woman—the flapper. In 1929 she traveled to Berlin, where she starred in *Tagebuch einer Verlorenen* (Diary of a Lost Girl, 1929) and *Die Büchse der Pandora* (Pandora's Box, 1930) for legendary director G. W. Pabst, as well as a French-language film, *Prix de Beauté* (Miss Europe, 1930), written by Pabst and René Clair, and directed by Augusto Genina.

80. Twenties film star Louise Brooks was the quintessential "flapper," known for her sleek dark "bob" and dramatic makeup. Library of Congress, Prints & Photographs Division, LC-DIG-ggbain-32453.

When Brooks returned to the United States, she refused Paramount's assignment to make a sound version of *The Canary Murder Case*, and was ostracized by the major Hollywood studios. She appeared in several minor roles in British talkies, some of them uncredited, before disappearing from the film scene in the late 1930s. Remarkably, during the 1950s, James Card, director of the George Eastman House museum in Rochester, recognized her at a film screening and discovered she was living in a small flat on a diet of gin, tea, and soda crackers. She had moved to Rochester to be able to see the silent films shown regularly at Eastman House. At Card's urging, she wrote several magazine stories and a memoir, *Lulu in Hollywood* (1982).

HARRY JOE BROWN (1890–1975), producer and director, received a law degree from Syracuse University in 1915. During the 1920s, Brown began producing silent Westerns and other action pictures for Monogram and for a time had his own production company. He directed and produced more than a dozen pictures starring Reed Howes (e.g., *The Bashful Buccaneer* [1925] and *The Dangerous Dude* [1926]), and is credited with producing 135 silent and talking pictures during a forty-year Hollywood career. Brown probably reached the pinnacle of his career in 1939 when, as associate producer to Daryll F. Zanuck on *Alexander's Ragtime Band*, he was nominated for a Best Picture Oscar. After World War II, Brown regularly produced Westerns starring Randolph Scott, including *Santa Fe* (1951).

JIM BROWN (1935–), actor, was an All-American student athlete at Syracuse University in football and lacrosse, who graduated in 1957. He went on to become one of the greatest running backs in National Football League history in a pro career with the Cleveland Browns. He made his screen debut in *Rio Concho* (1964), a Western starring Richard Boone and Tony Franciosa. He announced his retirement from football before the 1966 season, during the filming of *The Dirty Dozen* (1967), his first hit film. The six foot two, 232-pound Brown has since appeared in dozens of action films and television series.

PEARL S. BUCK (1892–1973), novelist, was a missionary to China, who, like many missionaries, returned to the United States to study agriculture at Cornell University. She received an M.A. degree in 1925. Buck won the Nobel Prize for Literature in 1938 for her 1937 novel, *The Good Earth*, which came to the screen that same year, starring Paul Muni and Luise Rainer. Other Buck novels that came to the screen include *Dragon Seed* (1942), which became a film by the same title in 1944, starring Katharine Hepburn and Walter Huston, and *Satan Never Sleeps* (1962), filmed the same year under the same title and starring William Holden and Clifton Webb.

81. Pearl S. Buck, ca. 1932. Photograph by Arnold Genthe. Library of Congress, Prints & Photographs Division, Arnold Genthe Collection: Negatives and Transparencies, LC-USZ62-10297.

BEN BURTT, JR. (1948–), sound engineer, who grew up in Jamesville (Onondaga County), is the son of a Syracuse University chemistry professor. After graduating from Nottingham High School in Syracuse, he studied physics at Allegheny College and earned a master's degree in film production at the University of Southern California. Fresh out of film school, he scored sound assignments with low-budget, high-profile filmmakers Roger Corman and Russ Meyer. A big break came in the mid-1970s when Burtt was hired by George Lucas as sound engineer for the original *Star Wars*. He won an Oscar for his work and has been in charge of sound for all six *Star Wars* films. He picked up a second Academy Award in 1989 for his work on *Indiana Jones: The Last Crusade*. Burtt joined with Pixar Animation Studio in 2005.

FRANCIS X. BUSHMAN (1883–1966), actor, a matinee idol who may have inspired the first use of that term, appeared in at least five Ithaca productions, including the first feature made in the city, *Dear Old Girl* (1913), which the Wharton Brothers produced for Essanay Pictures (before establishing their own studio). In that film, Bushman starred opposite his real-life wife, Beverly Bayne (1894–1982), as he did in most of his early pictures. While working in Ithaca, Bushman and Bayne resided on the Cornell campus in a home that now serves as the Alpha Phi sorority house. Bushman, whose chiseled good looks helped define American masculinity during the silent film era, was known as "king of the movies" during this period. His other Ithaca productions include *The Adopted Son* (1917), a Western shot in the wilds of South Hill, now home to Ithaca College.

JOHN CARPENTER (1948–), director, writer, producer, is a leading exponent of the horror film genre. *Halloween* (1978), which he wrote and directed, initiated a successful string of sequels and documentaries about making it that continued for decades. Carpenter was born and raised in Carthage (Jefferson County), the small North Country town that spawned silent-screen actress Lucille Hammill (*The Judgement House*, 1917) and 1940s ingénue Sally Bliss (*Meet Miss Bobby Socks*, 1944). While studying film at the University of Southern California, Carpenter worked on the *Resurrection of Bronco Billy*, which won the Oscar as best live-action short of 1970, and he directed *Dark Star*, the first of his science fiction films. Some of Carpenter's blockbuster hits are *Escape from New York* (1981), starring Kurt Russell; *Christine* (1983), about a Chrysler with a murderous mind of its own (and two sequels); and *Vampires* (1998), starring Daniel Baldwin.

RAYMOND CARVER (1938–1988), short story writer, taught in the Syracuse University creative writing program. His stories served as the source for more than a dozen films, including *Short Cuts* (1993) directed by Robert Altman, *Autumn of the Leaves* (Israel, 1995), and *Everything Goes* (2004). Tess Gallagher, Carver's wife and a Syracuse professor as well, collaborated with Carver and Altman on the screenplay for *Short Cuts*. The annual Raymond Carver Reading Series at Syracuse University brings poets and fiction writers to read their works aloud.

JOANNA CASSIDY (1945–), actress, from Haddonfield, New Jersey, studied art at Syracuse University during the late 1960s. She appeared in more than 125 feature films and television productions, including leading roles in *The Laughing Policeman* (1973), *Bank Shot* (1974), *The Late Show* (1977), and *Blade Runner* (1982).

IRENE CASTLE (1893–1969), a native of New Rochelle (Westchester County), came to Ithaca in 1916 to appear in *Patria*. She was already famous as a member of the dance team of Vernon and Irene Castle, and she was celebrated as one of America's best-dressed women. Her clothes, which she often designed for herself, and her chic bobbed hair put her at the cutting edge of style in the second decade of the twentieth century. To many, she embodied the new American woman, free of Victorian constraint, ready to be daring and playful. Vernon, her husband and partner, returned to his native Britain to join the Royal Flying Corps during World War I and lost his life. In 1923, Irene returned to Ithaca and married the second of her four husbands, Robert Treman, for whom Treman State Park is named. The couple's home on Cayuga Heights Road became Cornell University's Sigma Chi fraternity house.

GILBERT CATES (1934–), producer, director, born and raised in New York City, earned a bachelor's degree in 1955 and a master's degree in 1965 from Syracuse University. He launched his career as a film director with *I Never Sang for My Father* (1970), which earned three Academy Award nominations. Other Cates film credits include *Summer Wishes, Winter Dreams* (1993), starring Joanne Woodward and Sylvia Sydney, which he directed; and *Oh, God! Book II* (1980), starring George Burns, which he produced. He produced the annual Academy Awards presentation for television a dozen times between 1991 and 2006, for which he received seventeen Emmy awards. Dean of the UCLA School of Theater, Film and Television during the 1990s, he also served two terms as president of the Directors Guild of America. In 2006 he became producing director of the Geffen Playhouse in Los Angeles.

MARVIN J. CHOMSKY (1929–), director, producer, from New York City, studied speech and dramatic arts at Syracuse University, graduating in 1950. He is an enormously prolific and successful television director, with scores of made-for-TV movies and series episodes to his credit since 1962, including three Emmy-winning television efforts: *Holocaust* (1978), *Attica* (1980), and *Inside the Third Reich* (1982). Theatrical releases include *Murph the Surf* (1975), concerning the character behind a major political scandal of the administration of President Lyndon B. Johnson; *Macintosh and T. J.* (1975), Roy Rogers's last feature film; and *Good Luck, Miss Wyckoff* (1979), based on William Inge's novel about an interracial, intergenerational love affair set in 1956 Kansas.

DANE CLARK (1915–), actor, was born Bernard Zanville in Brooklyn and graduated from Cornell University in 1926 and St. John's University Law School in 1928. He boxed professionally and drove a truck during the Great Depression before turning to modeling and then acting, establishing himself as a quintessentially pugnacious tough guy. After starting out in radio and theater productions in New York, Clark made uncredited film appearances in two 1942 Hollywood features, *The Glass Key* and *The Pride of the Yankees*, and was billed as "Bernard Zanville" in *The Rear Gunner* (1943). Humphrey Bogart provided him with the name Dane Clark while the two worked together on *Action in the North Atlantic* (1943). A ubiquitous presence in World War II Warner Brothers action features, Clark went on to appear in more than 150 films and television programs. Some critics believe his greatest performance was as Danny Hawkins, the confused accidental killer in *Moonrise* (1948), which he made for Republic Pictures.

DICK CLARK (1929–), producer, actor, was born in Mount Vernon (Westchester County) and grew up in Utica. Attending Syracuse University, he got his first experience as a disc jockey on WAER-FM, the campus radio station. After graduating in 1951, he went to work for an uncle in the broadcasting business and made a connection to a Philadelphia station that needed to replace a disc jockey who had been handling its afternoon television teenage dance show. Clark got the job, and the TV show eventually went coast-to-coast as *American Bandstand,* running on the ABC network from 1963 to 1987. Clark was a major pop cultural entrepreneur for the next fifty years, specializing in producing television series and made-for-television movies with youth appeal. His main involvement with feature films came during the late 1960s, when the studios were looking for ways to capture the interest of young moviegoers. During 1968, Clark produced three films for theatrical release: *Killers Three,* a road film about ripping off a bootlegger and hitting the gas for California, which Clark also wrote and starred in; *Psych-Out* with Jack Nicholson, which takes place in San Francisco's Haight-Ashbury district during the Summer of Love; and *The Savage Seven,* about bikers out of control on an Indian reservation.

WALTER VAN TILBURG CLARK (1889–1971), fiction writer, taught English and drama and coached basketball and tennis at Cazenovia Central High School from 1936 to 1945. His novel *The Ox-bow Incident* (1940), a dark story about mob violence and the lynching of an innocent man, was adapted for the screen in a masterful script by Lamar Trotti. The 1943 film starred Henry Fonda and Dana Andrews and was directed by William Wellman. Clark also wrote the novel and the screenplay for *Track of the Cat* (1948), starring William Holden and Teresa Wright.

JACKIE COOGAN (1914–1984), actor, the son of vaudevillians, was born in a proverbial trunk but had deep roots in Syracuse, his father's hometown. Performing in his parents' act by the time he was four, Coogan often stayed at his grandfather's house in Syracuse between tours. Coogan's film career began when Charlie Chaplin discovered him in 1921, casting him in the title role of *The Kid* (1921). As a child actor during the silent period, Coogan appeared in such films as *Peck's Bad Boy* (1921), *Oliver Twist* (1922), and *A Boy of Flanders* (1924). In talkies, he played the title role in *Tom Sawyer* (1930) and reprised the role in *Huckleberry Finn* (1931). He appeared in more than 130 films and television series and is probably best remembered as Uncle Fester in *The Addams Family* (ABC series, 1964–66). Off screen, Coogan made headlines as well. Following his father's death in 1935, Coogan sued his mother and stepfather for a fair share of the millions he had made as a child actor but was awarded only $126,000. The case caused an

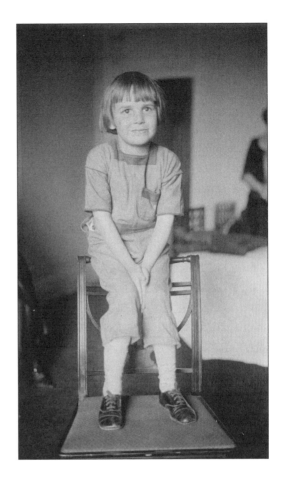

82. Jackie Coogan, child star during the silent movie era, was discovered by Charlie Chaplin in 1921. Library of Congress, Prints & Photographs Division, LC-DIG-ggbain-32235.

uproar and led to the California Legislature's passage of the Child Actors Bill, also known as the Coogan Act.

JAMES FENIMORE COOPER (1789–1851), fiction writer, thought by some critics to be America's first world-class novelist, was born in Burlington, New Jersey. His father, a member of Congress, had invested in the development of central New York land, and the family moved to Cooperstown in 1800, living in a mansion on Lake Otsego. Cooper attended the village school and then Yale, but he was expelled from college in his junior year for pranks, including bringing a donkey to class. Cooper later lived downstate in Mamaroneck in Westchester County, but critics agree that his most enduring and cinematic tales are set in the Mohawk Valley of his youth. *The Last of the Mohicans,* Cooper's most popular novel, was also his most popular film story, with some seventeen adaptations made over the years, including silent, sound, and television releases, and U.S., British, French, German, and Italian productions. A 1992 British production starring Daniel Day Lewis is arguably the most successful. *The Deerslayer* has come to the screen many times in many languages, beginning in the silent era with a U.S. production, *The Deerslayer* (1913), and a German adaptation, *Lederstrumpf* (1923), the latter directed by Arthur Wellin. Two sound-era productions of *The Deerslayer* were made in Hollywood, in 1943, starring Bruce Kellogg and Yvonne DeCarlo, and in 1957, with Lex Barker. Other Cooper works adapted for the screen include *The Pathfinder* (filmed in 1952 and 1996) and *The Leatherstocking Tales* (filmed in 1924 as a silent serial).

STEPHEN CRANE (1871–1900), fiction writer, the son of a Methodist minister in Port Jervis (Orange County), spent the spring semester of 1891 at Syracuse University, living in the Delta Upsilon house on Ostrom Avenue and captaining the varsity baseball team, for which he played catcher. Crane's great uncle, Jesse Peck, a founder of the university, had served as chancellor. *The Red Badge of Courage* (1951), based on Crane's most famous short story, starred Audie Murphy as a young Civil War soldier struggling to find his courage.

TOM CRUISE (1962–), actor, was born in Syracuse, the first of many towns in which he lived as a child and teenager, moving with his peripatetic family. He returned to Syracuse in his early teens and attended school at a Franciscan seminary before refocusing his career efforts toward acting. He made his debut in a bit part in director Franco Zeffirelli's *Endless Love* (1981) and went on to star in some of the highest grossing films of the 1980s, including *Top Gun* (1986) and *Rain Man* (1988). One of the highest paid performers in Hollywood, he has starred in such hits as *A Few Good Men* (1992), *Mission Impossible* (1996), *Minority Report* (2002), and *War of the Worlds* (2006).

DAN CURTIS (1927–2006), producer, director, writer, graduated from Syracuse University in 1950. He is best known for his daytime television series *Dark Shadows* (ABC, 1966–71), which was an attempt at making a hybrid of two genres that rarely cross paths: horror and soap opera. The series was actually based on a feature film, *House of Dark Shadows*, which Curtis wrote, directed, and produced for theatrical release. Another Curtis feature, *Burnt Offerings* (1976), also set in a haunted house, starred Karen Black and Oliver Reed. Curtis made dozens of movies and miniseries for television, including *Frankenstein* (1973), *The Winds of War* (1983), and a *Dark Shadows* revival movie (2004).

WILLIAM C. DAVIES (1932–), visual effects artist, was born in Auburn and graduated from Cayuga Community College in 1954. He was the uncredited airbrush artist on Stanley Kubrick's *2001: A Space Odyssey* (1968) and director of special effects for *The Doomsday Machine* (1972).

MAYA DEREN (1917–1961), filmmaker, was born in Kiev, Ukraine, during the Russian Revolution, escaping with her family from an anti-Semitic pogrom in 1922, and eventually settling in Syracuse, where her father was a staff psychiatrist at the New York State Institute for the Feeble-minded. She enrolled in Syracuse University to study journalism and political science, and was active in the Young People's Socialist League, where she met and married a fellow member, moving to New York City in 1935. It was not until 1941 that Deren picked up a camera for the first time. Her first film, *Meshes in the Afternoon* (1943), made with Alexander Hammid, her second husband, became an underground classic, which influenced such diverse filmmakers as Binghamton University professor Ken Jacobs (*Little Stabs at Happiness,* 1963) and David Lynch (*Lost Highway,* 1997). The film's success brought her a Guggenheim Fellowship, the first to be awarded to a filmmaker. Deren made eight films, including *The Witch's Cradle* (1944, a collaboration with Marcel Duchamp) and *The Very Eye of Night* (1958). She died of a brain hemorrhage at age forty-four. In 1986, the American Film Institute established the Maya Deren Award for independent filmmakers.

SCOTT "TAYE" DIGGS (1972–), actor, grew up in Rochester and studied musical theater at Syracuse University, graduating in 1993. Dubbed "the black Cary Grant" and "a modern-day Fred Astaire," Diggs emerged as a star on the Broadway stage before appearing in such films as *Chicago* (2002), *Drum* (South Africa, 2004), and *Rent* (2005).

BOB DISHY (1934–), actor, is a Brooklyn-born Syracuse University graduate (1955) who majored in drama. After making his start on Broadway, he went on to screen roles in *Damn Yankees* (1967), *The Tiger Makes Out* (1967), and *Lovers and Other Strangers* (1970). He is a frequent guest on television series and has appeared in many episodes of *Law and Order* as defense attorney Lawrence Weaver.

KIRK DOUGLAS (1915–), actor, was born Issur Danielovitch Demsky in Amsterdam (Montgomery County), to Russian Jewish immigrant parents. He attended St. Lawrence University in Canton, working as a waiter to put himself through school. An intercollegiate wrestler who dabbled in campus stage productions, he was elected president of his senior class, despite anti-Semitic protests from some alumni. After graduating in 1939, Douglas headed for Broadway, taking classes at the American Academy of Dramatic Art. He served in the navy during World War II, and when he returned, Lauren Bacall, a classmate at the academy, arranged a screen test for him, which launched his distinguished career as a film actor. Athletic and rugged, Douglas starred in such hits as *The Bad and the Beautiful* (1953), *Paths of Glory* (1957), *The Vikings* (1958), *Spartacus* (1960), *Seven Days in May* (1964), and *The Arrangement* (1969). Successfully transitioning to character roles with age, he has appeared in more than one hundred film and television productions. He is the author of seven books, including *The Ragman's Son: An Autobiography* (1988) and *My Stroke of Luck* (2002), concerning a stroke he suffered that partially debilitated his speech.

THEODORE DREISER (1871–1945), fiction writer, is a canonical figure in American literature. His novel *An American Tragedy* (1925) is based on the real-life case of Grace Brown, a factory worker in South Otselic (Chenango County) who was murdered by her employer, Chester Gillette of Cortland, at Big Moose Lake in the Adirondacks. Gillette's 1906 trial created a sensation. Dreiser's story was filmed as *A Place in the Sun* (1951), starring Montgomery Clift, Elizabeth Taylor, and Shelley Winters. A 1980 Filipino production in the Tagalog language, *Nakaw Na Pig-Ibig*, was likewise based on the Dreiser novel. Another Dreiser classic, *Sister Carrie*, came to the screen as *Carrie* (1952), starring Jennifer Jones, Lawrence Olivier, and Miriam Hopkins. (It has no relation to Brian DePalma's 1976 film of a Stephen King horror tale of the same title.)

MARY DUNCAN (1895–1993), actress, was a Virginian who graduated from Cornell University in 1916 and pursued a career on the Broadway stage. She began making films in 1927, appearing in the role of a vampish temptress in late-silent-era pictures, such as *Very Confidential* (1927) and *Soft Living* (1928), and made the transition to early sound productions, including *City Girl* (1930) and *Kismet* (1930). Her career abruptly ended in 1933 after she married international polo star (and heir to Bigelow–Sanford Carpet fortune) Laddie Sanford. Duncan lived to age ninety-eight as a philanthropist, entertaining the Duke and Duchess of Windsor, Rose Kennedy, and other high society figures at her Palm Beach and New York City homes.

DAN DURYEA (1907–1968), actor, from White Plains (Westchester County), graduated from Cornell in 1928 and pursued a career in advertising before turning to acting. He made his Broadway debut in 1935 as Leo, the weak, ineffectual son, in the original New York production of Lillian Hellman's classic play, *The Little Foxes*, and reprised the role in the screen adaptation a year later, thus launching his Hollywood career. Typecast as a cynical, sneering villain whom women found strangely fascinating, his most memorable roles were as the antagonist of Edward G. Robinson in two Fritz Lang films: *The Woman in the Window* (1944) and *Scarlet Street* (1945). He appeared in more than sixty films.

WALTER D. EDMONDS (1903–1998), fiction writer, was born in Boonville (Oneida County), and wrote stories set in the Mohawk Valley and along the Erie Canal. Among his novels adapted for film are *Rome Haul* (1935; see plate 16), filmed as *The Farmer Takes a Wife* (1935 and 1953); *Drums Along the Mohawk* (1936), filmed in 1939 (see plate 17); and *Chad Hanna* (1940; see plate 18), filmed the same year. Except for the 1953 remake of *The Farmer Takes a Wife*, all the film versions featured Henry Fonda. Fonda's ancestors were early Dutch settlers who founded the village of Fonda (Montgomery County) on the Mohawk River between Albany and Utica.

HENRY (1912–1992) and PHOEBE WOLKIND EPHRON (1914–1971), playwrights and screenwriters, were both born in New York City. He attended Syracuse University; she, Hunter College. Married in 1934, the Ephrons collaborated on stage plays, reaching Broadway in 1943 with *Three's a Family*, and adapting it for the screen the following year. The couple won the Best Screenplay Academy Award for *Captain Newman, M.D.* (1963) and were nominated for six Writers Guild Awards for their scripts, including those for three musicals: *Carousel* (1956), *Daddy Long Legs* (1955), and *There's No Business Like Show Business* (1954). Henry Ephron also served as producer for *Carousel* (1956), *Desk Set* (1957), and *A Certain Smile* (1958).

RAY EVANS (1915–2007), lyricist, was born in Salamanca (Cattaraugus County). He collaborated with Jay Livingston, a classmate at the University of Pennsylvania, on three Academy Award–winning songs: "Buttons and Bows" for *The Paleface* (1948) starring Bob Hope, "Mona Lisa" for *Captain Carey U.S.A.* (1950), and "Que Sera Sera" for Alfred Hitchcock's *The Man Who Knew Too Much* (1956). "Silver Bells," which they wrote for a 1951 Bob Hope comedy, *The Lemon Drop Kid*, has become a standard Christmas tune.

PETER FALK (1927–), actor, received a master's degree in public administration from Syracuse University's Maxwell School in 1953, and worked as an efficiency expert for the state of Connecticut's budget office. Bored with adding figures, he tried his hand at amateur dramatics, turning professional in 1955 with the encouragement of his teacher, Eva Le Gallienne. Falk received Oscar nominations as best supporting actor in *Murder, Inc.* (1960) and *Pocketful of Miracles* (1961), but is best remembered for his title role as the eccentric working-class detective in the long-running *Columbo* television series.

WALTER FARLEY (1915–1989), fiction writer, was born in Syracuse. His 1941 novel about a young boy's love for an Arabian horse, *The Black Stallion*, became a classic of children's literature and the basis for a series of children's films: *The Black*

Stallion (1979) starring Kelly Reno, Mickey Rooney, and Teri Garr; *The Black Stallion Returns* (1983); and *The Young Black Stallion* (2003), a prequel.

ANNETTE FUNICELLO (1942–), actress, was born in Utica. She made her show business debut as a Mouseketeer on the daily Disney afternoon television series, *The Mickey Mouse Club*, during the 1950s and parlayed her popularity as Annette (her original billing) into teen film stardom. She is best remembered for the series of beach movies she made with Frankie Avalon, beginning with *Beach Party* (1963). Funicello and Avalon reunited for *Back to the Beach*, a 1987 spoof of the genre.

ALLEN FUNT (1914–1999), producer and performer, graduated from Cornell University in 1935 with a bachelor's degree in fine arts. He is best known as creator, producer, and on-air host of the long-running *Candid Camera* television series, which Funt launched on radio as *Candid Microphone* in 1946. The *Candid Camera* premise—setting up absurd situations and documenting the unscripted reactions of unsuspecting people—has led some critics to credit Funt with inventing "reality television" decades before it became a popular television genre. Funt produced and starred in two Hollywood feature-length documentaries, *What Do You Say to a Naked Lady?* (1970) and *Money Talks* (1972), which include outtakes from the *Candid Camera* series deemed too risqué or grotesque for 1960s television.

FRANK GABRIELSON (1910–1980), screenwriter, graduated from Syracuse University in 1931. He got his start on Broadway as a performer before turning to screenwriting in the 1940s. His credits include such comedies as *Something for the Boys* (1944) with Carmen Miranda; *Don Juan Quilligan* (1945) with Joan Blondell and William Bendix; and *It Shouldn't Happen to a Dog* (1946) with Julie Andrews. He was a prolific television writer during the golden age of television drama, contributing scripts to such series as *Alfred Hitchcock Presents* and *The U.S. Steel Hour*. In 1960, he adapted L. Frank Baum's *Land of Oz* into a one-hour television script of the same name for *Shirley Temple's Storybook*. The production starred Temple, Jonathan Winters, Agnes Morehead, and Ben Blue. Other Gabrielson scripts for series episodes included "Hiawatha" and "Rumpelstiltskin."

RICHARD GERE (1949–), actor, was born in Philadelphia and grew up in the Syracuse area, graduating from North Syracuse High School in 1967, and winning a gymnastics scholarship to the University of Massachusetts, where he majored in philosophy. He began his career onstage in New York and London, making his film debut in *Report to the Commissioner* (1975). *Looking for Mr. Goodbar* (1977) was a breakout vehicle for Gere. He became one of Hollywood's most sought-after leading men, starring in such hits as *American Gigolo* (1980), *An Officer and a Gentleman* (1982), *Pretty Woman* (1990), and *The Hoax* (2006).

CARL GOTTLIEB (1938–), screenwriter, majored in speech and dramatic arts at Syracuse University, graduating in 1960. He is best known in film history for his collaboration with Robert Benchley on the script for Steven Spielberg's *Jaws* (1975), and as a screenwriter for the film's 1978 and 1983 sequels. But most of Gottlieb's work has been in comedy. His scripts include *The Jerk* (1979), starring Steve Martin, and *Doctor Detroit* (1983), starring Dan Aykroyd. Gottlieb also has many credits as an actor, director, and producer, mainly for television.

PETER GUBER (1942–), producer, is a Bostonian by birth and a 1964 Syracuse University graduate. His Guber-Peters production company, a partnership with Jon Peters, enjoyed great success in the eighties, with Guber producing such hits as *Flashdance* (1983), *The Color Purple* (1985), and *Rain Man* (1988). He headed Columbia Pictures from 1989 to 1994 and has since become an independent producer. Other Guber productions include *Midnight Express* (1978) and *Batman* (1989).

HOWARD HAWKS (1896–1977), director, studied engineering at Cornell before becoming one of Hollywood's leading *auteur* directors of screwball comedies and action pictures. Among his greatest hits are *Scarface* (1932), *Twentieth Century* (1934), *Bringing Up Baby* (1938), *Only Angels Have Wings* (1939), *The Big Sleep* (1946), *Red River* (1948; see plate 19), *Gentlemen Prefer Blondes* (1953), and *Rio Bravo* (1958; see plate 20). He directed forty-seven Hollywood features between 1930 and 1970. Remarkably, he was nominated for an Academy Award just once, for *Sergeant York* (1941; see plate 21). He received a career achievement Oscar as a master American filmmaker in 1974.

MIRIAM HOPKINS (1902–1972), actress, from Savannah, Georgia, attended Syracuse University in the early 1920s, but left college to become a dancer in New York City (she is mistakenly described as a Syracuse graduate in several film histories). Her films include *The Smiling Lieutenant* (1931) opposite Maurice Chevalier, *Dr. Jekyll and Mr. Hyde* (1932) with John Barrymore, and *The Old Maid* (1939) with Bette Davis.

PETER HYAMS (1943–), GEORGE-ANN SPOTA HYAMS, JOHN HYAMS (1972–), all graduates of Syracuse University, are a prominent film family. Peter is a producer, cinematographer, and screenwriter. Since graduating in 1964, he has worked as a television newsman and a professional drummer, and has painted several canvasses that hang in the Whitney Museum. However, he is best known as the director of more than twenty feature films, including *Capricorn One* (1978), *End of Days* (1999), and *Time Cop* (2007). George-Ann Spota who he met and married while the two were Syracuse classmates, is president of George Spota Productions, a company founded by her father. John Hyams, the couple's son, class of '93, is a director whose films include *One Dog Day* (1997) and *Rank* (2006).

CHARLES H. JOFFE (1929–), talent agent, producer, was born in Brooklyn and studied at Syracuse University, earning a B.S. degree in 1951. Joffe and partner Jack Rollins formed a talent agency that specialized in representing comedians, including Woody Allen, Billy Crystal, David Letterman, Tom Poston, and Robin Williams. In 1966, Joffe made his debut as a Broadway producer with Allen's play, *Don't Drink the Water,* and as a Hollywood film producer with Allen's *Take the Money and Run.* He has since produced or co-produced fifty features, including virtually all of Allen's films.

ALLAN JONES (1908–1992), actor, singer, was a son of a coal miner, working in the Pennsylvania mines as a boy. He won a scholarship to study music at Syracuse University, graduating in 1930. Jones became a popular romantic lead in such musicals as *Show Boat* (1936), opposite Irene Dunne, and *Firefly* (1937), opposite Jeanette MacDonald. After Zeppo Marx left the Marx Brothers, Jones played the male romantic lead in two Marx Brothers's comedies, *A Night at the Opera* (1935), opposite Kitty Carlisle, and *A Day at the Races* (1937), opposite Maureen O'Sullivan.

GRACE JONES (1952–), actress, model, singer, was born in Kingston, Jamaica, and studied theater at Syracuse University. The statuesque, flamboyant Jones was one of the more unforgettable characters to emerge from New York City's hedonistic Studio 54 disco scene during the late 1970s. (Studio 54 entrepreneur Steve Rubell was also a Syracuse graduate, class of 1965.) Jones's films include *Palmer's Pick-Up* (1999), *Shaka Zulu: The Citadel* (2001), *Wolf Girl* (2003), and *Shaka Zulu: The Last Great Warrior* (2005).

RAY JUNE (1895–1958), cinematographer, was born in Ithaca and attended Cornell University. He gained camera experience in the Army Signal Corps during World War I and, upon his discharge, took a job with the Wharton Brothers in his hometown as cameraman for *The Eagle's Eye* (1918), starring King Baggott. He subsequently worked on some 160 Hollywood productions, displaying meticulous craftsmanship in the cinematography of both major and minor films. During the 1930s, he was the highest paid cameraman in Hollywood, developing what became known as the "MGM look." Some of the studio's stars would balk at projects unless assured June was behind the camera, such was his reputation for making actors look good in close-ups. His final picture was *House Boat* (1958), starring Cary Grant and Sophia Loren.

MILTON KAHN (1934–), talent agent, was a Syracuse University graduate who founded Milton Kahn Associates, representing such stars as Gregory Peck, Joan Crawford, Steve Allen, Glenn Ford, Lee Grant, Chuck Norris, and Michael Landon; directors Roger Corman, Robert Aldrich, and Arthur Hiller; and musician Herb Alpert.

GARSON KANIN (1912–99), playwright, screenwriter, director, born in Rochester, started out as a vaudeville and burlesque comedian, and then directed on Broadway before shifting his attention to the movies. Often collaborating with Ruth Gordon, his wife, Kanin wrote more than twenty feature films, including such hilarious screenplays as *Tom, Dick and Harry* (1941, which he also directed) *Adam's Rib* (1949), *Pat and Mike* (1952), *The Rat Race* (1960, from his own play), and *The Live Wire* (1960).

MICHAEL KANIN (1910–1993), screenwriter, born in Rochester, began penning scripts for the movies in the late 1930s and won an Academy Award for his collaboration with Ring Lardner, Jr., on *Woman of the Year* (1942). Other notable scripts include *Centennial Summer* (1946), *When I Grow Up* (1951, which he also directed), and *The Outrage* (1964).

DORIS "DORIE" KENYON (1897–1979), actress, was born in Syracuse. Her father was a clergyman-poet. She got her start in films made by the Wharton Brothers, appearing in *The Great White Trail* (1917), a film whose production had to be moved to Saranac Lake when Ithaca uncharacteristically failed to deliver enough snow for a convincing Alaska setting. Kenyon appeared in dozens of silent films, playing leads opposite such stars as Thomas Meighan, Lewis Stone, and Milton Sills, who became the first of her four husbands in 1926. She made a smooth transition to sound and continued playing leads and supporting roles through the late thirties. She retired with credits in more than sixty films and television dramas.

NORMAN KERRY (1889–1956), actor, was a Rochester native who played dashing heroes as well as the occasional villain in late silent-era feature films. He often sported a fancy waxed mustache. Notable films included *The Hunchback of Notre Dame* (1923, in which he played Captain Phoebus), *The Spoilers* (1923), *The Phantom of the Opera* (1925), *Body and Soul* (1927), and *The Foreign Legion* (1928).

PHYLLIS KIRK (1929–), actress, was born in Syracuse. A former model, she played leads in Hollywood features during the 1950s. She was particularly popular as Nora Charles in the *Thin Man* television series. Her films include *A Life of Her Own* (1950), *The Iron Mistress* (1952), *House of Wax* (1953), *Johnny Concho* (1956), and *Back from Eternity* (1956).

FRANK LANGELLA (1940–), actor, came from Bayonne, New Jersey, to study drama at Syracuse University, graduating in 1959. He is perhaps best remembered in the title role of *Dracula* (1979). He also starred in Mel Brooks's *The Twelve Chairs* (1970), *Diary of a Mad Housewife* (1970), and *And God Created Woman* (1991). He played the role of CBS chair William Paley in *Goodnight and Good Luck* (2006), a biopic concerning newsman Edward R. Murrow.

ARTHUR LAURENTS (1918–), playwright and screenwriter, was born in Brooklyn and studied writing for the stage at Cornell University. His plays *Home of the Brave, The Time of the Cuckoo (Summertime), West Side Story,* and *Gypsy,* have all been adapted for the screen by others. Laurent has also written or collaborated on original screenplays for *Rope* (1948), *The Snake Pit* (1948), *Anna Lucasta* (1949), *Caught* (1949), *Anastasia* (1956), *Bonjour Tristesse* (1958), *The Way We Were* (1973, from his own novel) and *The Turning Point* (1977, which he also co-produced).

JERRY LEIDER (1932–), producer, is a Camden, New Jersey, native who graduated from Syracuse University in 1953, winning a Fulbright Fellowship that took him to Bristol University in England to continue his studies in drama. A versatile show business executive, Leider produced several successful shows on Broadway, did a stint as a programming executive for CBS television, and ran television operations for the Ashley talent agency. In 1969, he joined Warner Brothers, where he was instrumental in developing the made-for-television movie and then headed foreign production. Leider produced such films as *The Jazz Singer* (1980), *Morning Glory* (1993), *Dr. Jekyll and Ms. Hyde* (1995), and *Confessions of a Teenage Drama Queen* (2004). He occasionally returns to central New York as a guest lecturer at Syracuse University's Newhouse School of Public Communications.

SHELDON LEONARD (1907–1996), actor and producer, was born to an immigrant family in the Bronx, and studied drama and business administration at Syracuse University, graduating Phi Beta Kappa in 1929. A character actor whose "deez, dem, doze" accent was more recognizable to movie fans than his name, Leonard played Runyonesque gangsters in dozens of films for decades, including *Return of the Thin Man* (1934), *Tortilla Flat* (1942), *Guys and Dolls* (1955), and *Pocketful of Miracles* (1961). He gave a memorable performance in *It's A Wonder-*

ful Life (1946), playing both a friendly bartender and, in a dream sequence, the character's evil twin. Always in supporting roles on screen, he became a powerful figure behind the scenes in the television industry during the 1950s and 1960s, producing and directing such hit series as *I Spy, The Andy Griffith Show,* and *The Dick Van Dyke Show.*

WILLIAM LUNDIGAN (1914–1975), actor, was born and raised in Syracuse, attending Nottingham High School and Syracuse University, where he won varsity letters in basketball, football, and tennis, graduating in 1935. He started his career as a radio announcer at WFBL-Syracuse, whose studios were located at the Hotel Onondaga, which was owned by Lundigan's father. In a bizarre sequence of events, he was discovered by Universal production chief Charles R. Rogers, who heard Lundigan's voice while on a business trip and liked it for an overdubbing job he needed to cast. He had the studio bring Lundigan to New York City for a voice test, and when he saw the six foot two, 170-pound man behind the voice, sent him to Hollywood for a screen test. Lundigan appeared in more than eighty films and television series, often playing a somewhat bland leading man. His best-remembered films include *Dodge City* (1939), *The Old Maid* (1939), *The Sea Hawk* (1940), *Santa Fe Trail* (1940), *Pinky* (1949), *I'll Get By* (1950), and *The Way West* (1967). In later years he was spokesperson for the Chrysler Corporation.

ALISON LURIE (1926–), fiction writer, grew up in White Plains (Westchester County) and attended Radcliffe College. She joined the Cornell University English faculty in 1969 and later made children's literature a specialty. One of her best-known books, *The War Between the Tates* (novel, 1974; film, 1977), concerns the manner in which the wife (Elizabeth Ashley) of a professor (Richard Crenna) at mythical Corinth University deals with her husband's infidelity. *Foreign Affairs,* which won the Pulitzer Prize for Literature in 1985, came to the screen in 1993. It is also set at Corinth, describing the separate, unexpected sexual and romantic affairs of two faculty members during a sabbatical semester in England. *Imaginary Friends,* a 1967 novel concerning a group of people who believe they are in contact with aliens, was the source for a 1987 British television miniseries.

JOHN D. MACDONALD (1916–1986), fiction writer, one of the twentieth century's most prolific novelists, earned a degree in business administration at Syracuse University in 1938. He wrote scores of short stories and some seventy-seven mystery novels during his four-decade career, with his most popular character, private eye Travis McGee, "a knight in slightly tarnished armor," appearing in twenty-one books. Several of his novels have been adapted for the screen, most notably *The Executioner* (1958), which has twice been filmed as *Cape Fear.* The

1962 production starred Gregory Peck, Robert Mitchum, and Polly Bergen; the 1991 remake starred Robert DeNiro, Juliette Lewis, and Nick Nolte.

HAROLD MACGRATH (1871–1932), screenwriter, was a Syracusan who lived on James Street, within a few blocks of fellow writers Hervey Allen and Edward Noyes Westcott. Made deaf by a childhood accident, he worked as an editor for *The Syracuse Herald* when he began writing screenplays for the silent serials, *The Adventures of Kathlyn* (1914) and *The Million Dollar Mystery* (1914). His twenty-eight scripts included those for *Lure of the Mask* (1915), *The Ragged Edge* (1923), and *Man on the Box* (1925).

GAVIN MACLEOD (1930–), actor, was born Allan See in Mount Kisco (Westchester County) and raised in nearby Pleasantville. He studied acting at Ithaca College, graduating in 1952, and debuted on Broadway in *A Hatful of Rain* in 1956. MacLeod, who took his stage name from his Ithaca drama professor Beatrice MacLeod, is mainly known for his television roles as Murray on *The Mary Tyler Moore Show* and Captain Stubing on *The Love Boat*. He appeared in supporting roles in a half dozen feature films, notably Blake Edwards's *Operation Petticoat* (1959), starring Cary Grant and Tony Curtis, and *The Comic* (1969), starring Dick Van Dyke as a silent-era comedian.

GORDON MACRAE (1921–1986), actor, was born in East Orange, New Jersey, and raised in Syracuse, attending the Charles Andrews School and Nottingham High School, where he spent much of his time singing and acting in the drama club, of which he was a charter member. As a young teenager, he performed in the Syracuse University Children's Theater. At nineteen, he won a talent contest whose prize was a trip to the New York World's Fair to perform with the Harry James Orchestra. He remained downstate, finding work as a band singer and in radio. In 1948 he signed a seven-year contract with Warner Brothers that put him in a string of musicals, including *Look for the Silver Lining* (1949) with June Haver and Ray Bolger, and *Tea for Two* (1950), the first of five films he made with Doris Day. He gave his best and most well-known performances in two Rogers and Hammerstein musicals in which he costarred with Shirley Jones: *Oklahoma* (1955) and *Carousel* (1956). His career was interrupted by a long struggle with alcoholism, after which he performed principally in television.

ALBERT MAYSLES (1933–), documentary filmmaker, was a Bostonian by birth who studied psychology at Syracuse University, graduating in 1949. His first film, *Psychiatry in Russia* (1955), marked his transition to filmmaking. In partnership with his brother David (1931–87), he emerged as a leading exponent of *cinema verité*. Among the most celebrated of their thirty-six films are *Salesman* (1969), *Monterey Pop* (1969), *Gimme Shelter* (1970), and *Grey Gardens* (1975). Maysles has continued to make films since his brother's death, including documentaries on the Dalai Lama and the work of grand-scale artists Christo and Jeanne-Claude; and a self-portrait, *Handheld and From the Heart* (forthcoming).

EDIE MCCLURG (1951–), actor, became interested in radio as an undergraduate at the University of Missouri in her native Kansas City, pursuing a career as music librarian, on-air personality, and station manager at KCUR, the school's NPR affiliate. She came east to earn a master's degree in television, radio, and film at Syracuse University in 1970 and shifted her career focus to screen acting soon after. Since her debut in *Carrie* (1976), she has appeared in more than 140 films (e.g., *Ferris Buehler's Day Off*, 1986), television programs (e.g., *CSI*), and cartoon voiceover roles (e.g., *The Little Mermaid* animated films, 1989 and 2000). She typically plays secretaries, salespeople, bureaucrats, and other "plain folk" character parts.

JAY MCINERNEY (1955–), fiction writer, earned a master of fine arts degree in 1984 from the Syracuse University Creative Writing Program, studying with Raymond Carver and Tobias Wolff. He completed requirements for the program by submitting the manuscript for his novel, *Bright Lights, Big City*. McInerney then adapted the novel into the screenplay for the 1988 picture of the same name, starring Michael J. Fox as a dissolute writer. The writer collaborated on an original screenplay for *Gia* (1998), a made-for-TV biopic starring Angelina Jolie as fashion model Gia Carangi. A McInerney short story was the source for *Smoke Screens*, a 1997 independent film.

ADOLPHE MENJOU (1890–1963), actor, dreamed of a show business career in childhood, but his father opposed the idea, forcing him to enroll in the engineering program at Cornell University. As a student, he defiantly took courses in the liberal arts, and then dropped out of school in his junior year. He moved to New York City in 1915 hoping to break into the theater, but instead found bit parts in the city's thriving silent-era film industry at the Vitagraph, Biograph, and Edison studios. After years of mostly uncredited roles, he got his break with a featured character part in *The Faith Healer* (1921). During the sound era, he emerged as a quintessentially dapper and debonair gentleman in sophisticated drawing-room comedies, a leading man during the 1930s and 1940s, and a character actor thereafter. He was fluent in six languages and was known as one of Hollywood's best-dressed men. He was nominated for a Best Actor Academy Award for his role in *The Front Page* (1931), costarring Pat O'Brien. The film was an adaptation of the Ben Hecht–Charles MacArthur newsroom play later made as *His Girl Friday* (1939). Menjou's 150 credits include roles in *The Sheik* (1921) with Rudolph Valentino, *Morocco* (1930) with Marlene

Dietrich and Gary Cooper, Josef von Sternberg's *Dancing in the Dark* (1949) with William Powell, and *The Ambassador's Daughter* (1956) with Olivia DeHavilland.

GLEN MORGAN (1960–), television writer and producer, was born in Syracuse and grew up in the Fairmount section. He met his writing partner, James Wong, while the two were students at Loyola Marymount University in Los Angeles, and they broke into the business with their screenplay for *The Boys Next Door* (1985), a horror film starring Charlie Sheen. Morgan's other features include *Final Destination* (2000), which Wong directed, and *Black Christmas* (2006), a spoof of a slasher film. But Morgan and Wong are mainly known as prolific writers of episodic television, penning multiple scripts for such series as *The X-Files*, *Millennium*, and the 2007 revival of *The Bionic Woman*. Morgan produced the 2003 remake of *Willard*.

JAMES MORRIS (1950–), producer, graduated from Syracuse University in 1971 with a bachelor's degree in speech and dramatic arts. He served as president of Industrial Light and Magic, a special-effects company formed by George Lucas that has done special effects for such blockbusters as *The Empire Strikes Back* (1980), *E.T.* (1982), *The Return of the Jedi* (1983), *Indiana Jones and the Temple of Doom* (1984), *Back to the Future* (1985), *Terminator 2* (1991), and *Jurassic Park* (1993). He then became production chief at Pixar Animation Studios.

VIGGO MORTENSEN (1958–), actor, was born in Watertown (Jefferson County) and graduated from St. Lawrence University in 1980. Since his screen debut as a young Amish farmer in Peter Weir's *Witness* (1985), he has appeared in more than forty films and television productions. He is probably best known for his role as Aragorn in the *Lord of the Rings* trilogy (2001–3). He also wrote the song that he sings in *The Return of the King*, the final installment of the trilogy. Other Mortensen roles include Tom Stall in *A History of Violence* (2005) and the title role in *Alariste* (2006). He was awarded an honorary degree by St. Lawrence University in 2006.

VLADIMIR NABOKOV (1899–1977), fiction writer, taught Russian literature at Cornell University from 1948 to 1959. While living in a house on East Seneca Street in Ithaca, one of the city streets connecting downtown and Collegetown, the Russian-born novelist reportedly rode Ithaca city busses in search of American vernacular dialogue for his novel, *Lolita*, which came to the screen in 1962 starring James Mason and Shelley Winters. Cornell serves as the model for Wordsmith University, the setting of several of his novels.

JOHN NICHOLS (1940–), fiction writer, graduated from Hamilton College in 1962. Several years later, Alan Pakula brought a Nichols story to the screen in his directorial debut, *The Sterile Cuckoo* (1969). Starring Liza Minnelli, it was filmed in several locations in central New York, and local students were used onscreen. The premiere was held at Syracuse University, with Pakula present.

GEORGE WILBUR PECK (1840–1916), writer and politician, was born in Henderson, near Sackett's Harbor, on the eastern shore of Lake Ontario. A newspaper publisher and politician (he was elected governor of Wisconsin in 1890), Peck enjoyed writing short stories as a sideline. His popular character William Peck, an incorrigibly mischievous young boy, served as the source for his Peck series of humorous sketches, initially published in his newspaper and later collected into books. After Peck's death, his stories were made into several films: *Peck's Bad Boy* (1921) starring Jackie Coogan; a talkie remake of *Peck's Bad Boy* (1934) starring Jackie Cooper; and *Peck's Bad Boy at the Circus* (1938). The character also inspired *Peck's Bad Girl* (1921) starring Mabel Normand and a 1959 remake of that film. "Peck's bad boy" became a phrase synonymous with disobedient children.

ERNEST PINTOFF (1931–2002), director and animator, was originally from Watertown, Connecticut. After graduating from Syracuse University in 1953, he joined United Productions of America (UPA) as an animator, and eventually set up his own studio. A highly original comic cartoonist, he turned out many entertaining animated shorts for theatrical and television release and did many TV commercials as well. He was nominated for an Academy Award for best animated short in 1959 for "The Violinist" and won the Oscar in 1963 for "The Critic" (1963), a satire on modern art written and narrated by Mel Brooks. Other Pintoff gems include "The Interview" (1961) and "The Old Man and the Flower" (1962). With the market for artistic animated shorts small, Pintoff usually paid the rent directing episodic television.

SUZANNE PLESHETTE (1937–2008), actor, grew up in New York City, where her father was the manager of the Paramount and Brooklyn Paramount theaters during the big band era. She graduated from Syracuse University in 1958. An attractive and intelligent leading lady on stage, screen, and television, she appeared in more than one hundred productions, mostly in television but including such feature films as Alfred Hitchcock's *The Birds* (1963), *If This is Tuesday, It Must be Belgium* (1969), and *Support Your Local Gunfighter* (1971).

WILLIAM PRINCE (1913–), actor, was born in Nichols (Tioga County) and educated at Cornell University. He began as a stage actor in the 1930s and played leads and supporting roles in Hollywood films, including *Cyrano de Bergerac* (1950), *Sacco and Vanzetti* (1971), *Blade* (1973), and *Network* (1976).

ALAN RAFKIN (1928–2001), television director and producer, was born in New York City. After graduating from Syracuse University in 1950, he worked as a nightclub comic and actor, finally focusing his efforts at directing. Working in television, he directed episodes for some seventy-five sitcoms, including such top-rated series as *The Andy Griffith Show*, *Bewitched*, *The Mary Tyler Moore Show*, *Sanford and Son*, *M*A*S*H*, *Laverne and Shirley*, and *Murphy Brown*. His papers, including more than seven hundred shooting scripts as well as various video memorabilia, are held in the special collections of Syracuse University Library.

MARTIN RANSOHOFF (1927–), producer, from New Orleans, attended school at Colgate University in Hamilton. He founded Filmways, a major television production company, and produced some thirty-five films, including *The Americanization of Emily* (1964) with Julie Andrews, *Topkapi* (1964) with Peter Ustinov, *The Sandpiper* with Elizabeth Taylor and Richard Burton (1965, for which Ransohoff also wrote the original story), *The Loved One* (1965) directed by Tony Richardson, and *The Wanderers* (1979), an adaptation of the novel about Bronx street life by Cornell graduate Richard Price.

NICHOLAS RAY (1911–1979), director, a dynamic, socially conscious filmmaker with a gift for achieving fluid motion on the screen, was long the darling of French *auteur* critics when he arrived at Binghamton University in 1971 to teach cinema. Refusing to offer traditional classes, Ray began production of a film, *We Can't Go Home Again* (1976), and simply invited students to collaborate in the making of it. Performance artist Leslie Levinson (*The King of Comedy*, 1983) makes one of her rare screen appearances in a featured role. Ray, who directed such Hollywood classics as *Johnny Guitar* (1954), starring Joan Crawford, and *Rebel Without A Cause* (1955), starring James Dean, plays himself in the Binghamton production.

PHILIP REED (1908–1996), actor, was born Milton LeRoy Treinis in New York City and educated at Cornell. After some experience in vaudeville and on the legitimate stage, Reed made his film debut in *College Coach* (1933), starring Dick Powell and directed by William Wellman. Tall, dark, and athletically handsome, he was a leading man and second lead in more than fifty Hollywood feature films, most of them undistinguished. He sometimes played self-centered characters with mean streaks, as in *Weekend For Three* (1941), starring Jane Wyatt, in which he plays a vain former boyfriend who almost wrecks a marriage.

CHRISTOPHER REEVE (1952–2004), actor, was born in New York City and raised in Princeton, New Jersey. An acting prodigy, he toured as leading man opposite Celeste Holm in *The Irregular Verb to Love* after graduating from high school. An outstanding student at Cornell University, he was one of two applicants accepted to the Juilliard School's graduate acting program with advanced standing in lieu of a bachelor's degree (Robin Williams was the other). After making his first credited screen appearances in minor roles in *Enemies* (1974) and *Gray Lady Down* (1976), Reeve burst into stardom with the title role in *Superman* (1978), and made three sequels during the next nine years. He continued to appear in a wide variety of screen roles, including dramas (e.g., *The Bostonians*, 1984), comedies (e.g., *Noises Off*, 1992), and thrillers (e.g., *Above Suspicion*, 1995). After sustaining injuries in a 1995 horseback riding accident that paralyzed him from the neck down, he became an advocate—and in many cases, an inspiration—for the physically disabled. He served as narrator for *Without Pity: A Film About Abilities* (1996), a documentary made for HBO, and appeared in several television roles before his death.

HAL ROACH (1892–1992), producer, was born in Elmira (Chemung County). His father was a jeweler and his mother ran the Roach home as a boarding house, a circumstance that motivated Roach to leave as a teenager. He tried his fortune at gold prospecting in Alaska and construction work in the Mojave Desert before discovering he could use his Stetson and boots to get work as a film extra in Hollywood. By 1918, he was producing films with the silent cinema's greatest comedians, including Harold Lloyd, Charlie Chase, and the team of Stan Laurel and Oliver Hardy. He is probably best known for the *Our Gang* comedies (a series of short films, later repackaged for television as *The Little Rascals*). During the television era, he produced a long list of successful sitcoms, including *My Little Margie*, *The Life of Riley*, and *The Amos 'n Andy Show*. Living to age one hundred, Roach was active in the cable era, producing programs for the Disney Channel.

KENNETH LEWIS ROBERTS (1885–1957), fiction writer from Kennebunk, Maine, was a 1908 Cornell University graduate. His short story, "Good Will and Almond Shells," was adapted for the silent screen as *The Shell Game* (1918), directed by George Baker. Roberts's most famous novel, *Northwest Passage* (1937), concerns the eighteenth-century efforts of the Robert Rogers expedition to discover a water route across North America to the Pacific. It is set along New York's Mohawk River, the only nonmountainous route from New England west to the Mississippi Valley. (It would subsequently become the route of the Erie Canal, the New York Central Railroad, and the New York State Thruway.) Released in 1940, the film starred Spencer Tracy as Rogers and was directed by King Vidor. A television series based on *Northwest Passage* was later re-edited into another feature film, *Fury River* (1961), for international release. Other Roberts novels gave title to such films as *Captain Caution* (1940), set in the war of 1812, starring Victor Mature; and *Lydia Bailey* (1952), set in the Caribbean during the colonial period, starring Anne Francis in the title role.

MORGAN ANDREW ROBERTSON (1861–1915), fiction writer, was born in the Lake Ontario port city of Oswego, the son of a Great Lakes ship captain. Robertson wrote more than two hundred sea tales. Three silent-era features were based on Robertson's work. *Masters of Men* (1923), with Earle Williams, is based on a Robertson novel set in Cuba. *The Closing of the Circuit* and *The Enemies*, both made in 1915, were adapted from two of his short stories.

CASEY ROBINSON (1903–1979), from Logan, Utah, was a 1927 Cornell graduate who became a prolific screenwriter, with credits in more than seventy features. He was principally associated with Warner Brothers, writing such films as *Captain Blood* (1935), starring Errol Flynn, which brought him an Academy Award nomination, *Four's a Crowd* (1938), and *Passage to Marseilles* (1944), starring Humphrey Bogart. Robinson was particularly effective in writing acerbic one-liners for Bette Davis, and he was assigned to script many of her pictures, including *Dark Victory* (1939), *Now, Voyager* (1942), and *The Corn is Green* (1945). He was less successful in producing and directing efforts.

ANDY ROONEY (1919–), writer, was born in Albany and graduated from Colgate University in 1940. Although best known for his commentaries delivered on the CBS News weekly television series *60 Minutes*, Rooney began his career in 1946 as a screenwriter at MGM, and later wrote for individual performers, including Arthur Godfrey, Garry Moore, and Victor Borge. He wrote and produced documentaries for television, including *Lost, Stolen or Strayed* (1969), which won an Emmy.

CARL SAGAN (1934–1996), fiction writer and documentarian, was also an astronomer and member of the Cornell University faculty. *Contact*, his 1985 science fiction novel, was adapted for the screen in 1997 by Robert Zemeckis. Sagan also produced and narrated the PBS series *Cosmos*. One of public television's most popular science series, it made Sagan a celebrity. His catchphrase, "billions and billions," became part of common parlance.

GENE SAKS (1921–), actor and director, attended Cornell University and trained for the stage at the Drama Workshop of the New School for Social Research in New York City. Making his Broadway debut in the original production of *South Pacific*, he appeared in many plays, both on Broadway and off, and directed many of Broadway's biggest hits. He directed such films as *Barefoot in the Park* (1967), *The Odd Couple* (1968), *Cactus Flower* (1969), *The Last of the Red Hot Lovers* (1972), *Mame* (1974), and *Brighton Beach Memoirs* (1986). He has appeared in many others, including *A Thousand Clowns* (1965), *The Prisoner of Second Avenue* (1975), *The Goodbye People* (1984), and Woody Allen's *Deconstructing Harry* (1997).

TOM EVERETT SCOTT (1970–), actor, from East Bridgewater, Massachusetts, studied drama at Syracuse University, receiving his bachelor's degree in 1992. Scott has since appeared in more than thirty films and television productions. He played the role of Bert Cates, the biology teacher on trial for teaching evolution, in a remake of *Inherit the Wind* (1999), starring Jack Lemmon and George C. Scott. Other feature-film credits include *An American Werewolf in Paris* (1997), *Dead Man on Campus* (1998), *Sexual Life* (2005), and *Because I Said So* (2007), starring Diane Keaton.

HENRY SELICK (1952–), writer, director, producer, and visual effects artist, was originally from Glen Ridge, New Jersey, and graduated in 1974 from Syracuse University's College of Visual and Performing Arts. Selick is the artist behind the animated feature *The Nightmare before Christmas* (1993), and such animated shorts as *Slow Bob in the Lower Dimensions* (1991) and *Moongirl* (2005). As a director, he has shown a mastery of special effects in such films as *Monkeybone* (2001) and *James and the Giant Peach* (1996).

ROD SERLING (1924–1975), screenwriter and producer, was born in Syracuse and raised in Binghamton. He began writing during the golden age of live drama presentations on television, with his teleplay "Patterns" winning high critical praise and five Emmy Awards. It was adapted as a feature film in 1956, starring Van Heflin. Serling reached his peak of success—and became a household face as on-air host—with his long-running *Twilight Zone* series, an anthology of sci-fi and supernatural-themed dramas. Serling screenplays include *Requiem for a Heavyweight* (1962), which he adapted from his own teleplay; *Seven Days in May* (1964); and *Planet of the Apes* (1968). Serling named his studio Cayuga Productions, Inc., and many *Twilight Zone* stories are set in upstate New York, including a story about a confused woman stuck in the Binghamton Greyhound terminal and another about a Manhattan corporate executive who yearns to return to Willoughby, the small upstate town of his youth, and suddenly finds himself there. He also held a special premiere of the 1983 *Twilight Zone* feature film at the Crest Theatre in Binghamton. Serling's papers are held by Ithaca College, where he was a faculty member.

ALLAN J. SHALLECK (1929–2006), writer of children's literature and writer-director of animated films, studied speech and dramatic arts at Syracuse University, graduating in 1950. He collaborated with Margret Rey, one of the two original creators of the Curious George children's book character, to bring the mischievous monkey to television in an animated cartoon series. The pair later collaborated on more than two dozen books about George. Shalleck wrote and directed more than one hundred episodes for the Disney Channel's *Curious George* series, as well as the

83. Actress Norma Shearer Thalberg after her marriage, with her husband, MGM studio production chief Irving Thalberg, July 24, 1929. Library of Congress, Prints & Photographs Division, LC-DIG-npcc-17843.

animated feature film, *Curious George* (2006), whose all-star cast of voices included Will Ferrell, Drew Barrymore, and Dick Van Dyke.

MELVILLE SHAVELSON (1917–), screenwriter, director, and producer, is a native Brooklynite who graduated from Cornell University in 1939. While working as a press agent, he teamed with Milt Josefsberg to write comedy for Bob Hope, who had a weekly radio program. They followed Hope into the movies, helping him create his signature comic persona in such films as *The Princess and the Pirate* (1944) and *The Great Lover* (1949). During the 1950s, Shavelson wrote and directed dramatic vehicles for Hope, such as *The Seven Little Foys* (1955) and *Beau James* (1957). Shavelson wrote screenplays for more than forty features and directed twenty.

NORMA SHEARER (1900–1983), an actress known as one of the leading ladies of the silent screen, began appearing onscreen in 1920 and had her first starring role in *A Clouded Name* (1923), a film made in Syracuse. The principle shooting location was the Calthrop mansion on the 3500 block of South Salina Street. Its success would lead to Shearer's MGM contract. She eventually married the studio's production chief, Irving Thalberg. She appeared in some sixty films.

LEE SHUBERT (1873–1953), SAM SHUBERT (1875–1905), and JACOB J. SHUBERT (1879–1963), brothers and theater owners, were immigrants to the United

States from Eastern Europe. The Shubert family arrived in Syracuse in 1882, and the brothers began their careers selling newspapers in front of the Weiting Opera House. They gradually worked their ways indoors, with J. J. running the Weiting, Sam managing the nearby Bastable Theatre, and Lee serving as treasurer for both. After Sam's death in a train accident in 1905, Lee and J. J. went on to New York City, where they built the largest theatrical circuit in the country.

SIME SILVERMAN (1873–1933), journalist, was a Cortland (Cortland County) native who became a newspaperman with *The New York Morning Telegraph*. In 1905, he founded a weekly magazine devoted exclusively to show business content, which he called *Variety*. An instant success, it survives today as an important communications organ in the show business world, recognized for its trend-setting style and vernacular and its authority in the business end of show business. Silverman served as editor and publisher for many years, and is credited with writing the famous *Variety* headline, "Wall Street Lays an Egg," to announce the stock market crash of 1929. He launched a daily Hollywood edition of the paper shortly before his death.

WILLIAM SMITH (1932–), actor, began appearing as an extra during the 1940s in such films as *Meet Me in St. Louis* (1944) with Judy Garland and *Going My Way* (1945) with Bing Crosby when he was a child growing up in Los Angeles. Smith has appeared in some three hundred films and television shows during his career, playing bikers, thugs, cowboys and bare-fisted tough guys of every description. He attended Syracuse University in the 1950s after returning from air force service in Korea, and graduated cum laude at UCLA. A small sampling of his feature credits includes *High School Confidential* (1958), *Three Guns for Texas* (1964), *Blood and Guts* (1978), *Conan the Barbarian* (1982), *Interview with a Zombie* (1995), and *Her Morbid Desires* (2007).

JIMMY SMITS (1955–), actor, is a Brooklyn native and graduate of Brooklyn College who came to Ithaca to study acting at Cornell University, earning an MFA degree in 1982. The following year, he played Hamlet in a New York Shakespeare Festival production at the Public Theater. Most widely known for his roles in television series, notably *L.A. Law*, *NYPD Blue*, and *The West Wing*, Smits also has a dozen film credits, including starring roles in *Bless the Child* (2000), a thriller costarring Kim Basinger, and *Price of Glory* (2000), in which he plays a washed-up boxer; and a supporting role in *Star Wars Episode II: Attack of the Clones* (2002).

AARON SORKIN (1961–), playwright, screenwriter, and television producer, grew up in Scarsdale (Westchester County) and studied theater at Syracuse University, earning a bachelor of fine arts degree in 1981. He adapted his stage play into the hit film *A Few Good Men* (1992), starring Jack Nicholson and Tom Cruise, and followed with screenplays for *Malice* (1993), starring Alec Baldwin and Nicole Kidman, and *The American President* (1995). He is best known as the creator and writer of the critically acclaimed hit TV series, *The West Wing*. Sorkin sponsors an annual week-long trip to his Los Angeles production studio for Syracuse University acting and filmmaking students.

JERRY STILLER (1927–), actor, was a drama major at Syracuse University, and got some of his first experiences as a stand-up comedian in downtown Syracuse before his graduation in 1950. Stiller and wife Anne Mearer toured as a comedy team for many years, appearing on Ed Sullivan's TV variety series thirty-six times. Best known for his television sitcom roles in *Seinfeld* and *The King of Queens*, Stiller has appeared in many film comedies, including *Airport '75* (1975), *The Ritz* (1976), *Hairspray* (1988 and the 2007 remake), and *Zoolander* (2001), one of a dozen he has appeared in with his son, Ben Stiller. Father and son constituted the entire cast of *Shoeshine* (1987), a ten-minute film that won the grand prize for shorts at the Montreal Film Festival and was nominated for an Academy Award.

NORMA TALMADGE (1897–1957), actress (and sister of screen actresses Constance and Natalie), was among the greatest stars of the silent era. She began her film career in 1910 at the Vitagraph Studios in Brooklyn, a short streetcar ride from her home. In 1916, she met and married exhibitor Joseph M. Schenck. The couple formed the Norma Talmadge Film Corporation, which became one of the most lucrative partnerships in show business. Talmadge starred in *The Secret of Storm Country* (1918), based on the 1917 novel by Ithacan Grace Miller White. Of the many films produced in Ithaca during the silent era, it was the only one whose story was set in central New York.

FRANCHOT TONE (1905–68), actor, was born in the city of Niagara Falls, the son of a socially prominent industrialist. He was president of the Dramatic Society while a student at Cornell University, graduating in 1927. A star of stage and screen, he was generally typecast as a playboy or successful man-about-town, roles compatible with his well-bred looks and manners. He was nominated for an Academy Award for his performance in *Mutiny on the Bounty* (1935) and appeared in such notable films as *The Lives of a Bengal Lancer* (1935), *Three Comrades* (1939), *Five Graves to Cairo* (1943), *Phantom Lady* (1944), and *Advise and Consent* (1962).

DOROTHY TREE (1908–), actress, born Dorothy Triebitz, came from Brooklyn to study at Cornell University. She began her career on the Broadway stage, making her film debut in *Just Imagine* (1930), a musical comedy. A founding member of the Screen Actors Guild, she played leads, second leads, and supporting roles in Hollywood films, including *Bridge of Sighs* (1936), *Abe Lincoln in Illinois* (1940), *Knute Rockne—All American* (1940), *The Asphalt Jungle* (1950), and *The Men* (1950). A political activist with strong feminist views, she left the film world out of disgust during the McCarthy red hunts, changed her named to Dorthy Uris, and began a second successful career as a speech and voice coach for the Metropolitan Opera. She is the author of several books, including *To Sing in English* (1971), a textbook still in wide use.

MARK TWAIN (1835–1910), fiction writer, journalist, and essayist, born Samuel Clemens, is usually associated with life along the Mississippi River, the setting for his most beloved stories. But the Missouri-born Twain spent a good portion of his life (and, as he might say, much of his death as well) in New York State. In 1869, Twain bought a one-third interest in the *Buffalo Express* and served as an editor and a columnist for the newspaper. Soon after, he married Olivia Langdon, the daughter of a wealthy coal baron. When the Buffalo newspaper failed, the couple moved to the Langdon family's home, Quarry Farm, in Elmira. Although they took up residence in Hartford several years later, Twain returned to Elmira often, writing some of his most famous novels at his study at Quarry Farm, overlooking Elmira and the Chemung Valley. In an 1874 letter to William Dean Howells, he described it as "the loveliest study you ever saw, octagonal with a peak roof, each filled with a spacious window . . .

84. Stereograph of writer Mark Twain (Samuel Langhorne Clemens), ca. 1907. Library of Congress, Prints & Photographs Division, LC-USZ62-127284.

perched on the top of an elevation that commands leagues of valley and city and retreating ranges of distant blue hills." Twain's papers and many personal effects are held by Elmira College's Center for Mark Twain Studies, and the author is buried in the city's Woodlawn Cemetery. Many films have been made of Twain's fiction, especially his most popular novels, which have been remade many times over. *The Adventures of Tom Sawyer* (1876) was the subject of two silent productions (1907, 1917), five sound features for theatrical release (1930, 1938, 1947, 1984, 1995, 2000), and a handful of made-for-TV movies and series. Adaptations of the *Adventures of Huckleberry Finn* (1884) have starred Jackie Coogan (1931) and Mickey Rooney (1939). Other films of Twain fiction include *A Connecticut Yankee in King Arthur's Court* (1949) with Bing Crosby; *The Prince and the Pauper* (1937) with Errol Flynn and Claude Rains; and *The Man with a Million* (1954, based on "The $1,000,000 Banknote") with Gregory Peck. A biopic, *Adventures of Mark Twain* (1944), starred Fredric March in the title role.

OLIVER A. UNGER (1914–1981), producer, attended Syracuse University, graduating in 1935. The son of a well-to-do Chicago family, he took a position as vice president of Hoffberg Productions, a film distributor and production house specializing in educational material. He remained at or near the top for the rest of his career, which included a stint as head of the entertainment division of Commonwealth United Films. He formed his own company in the 1960s and produced such films as *Mozambique* (1965), *Ten Little Indians* (1965), *The Face of Fu Manchu* (1965), and *Our Man in Marrakesh* (1966), but remained associated with Commonwealth United for the balance of his career.

JIMMIE VAN HEUSEN (1913–1990), songwriter, was born Chet Babcock. A Syracuse native, he began writing songs as a high school student and, at age sixteen, became one of the first radio disc jockeys in the area. He briefly attended Cazenovia Seminary (now Cazenovia College) and then Syracuse University, where he studied music with Howard Lyman. While at the university, he befriended Jerry Arlen, the brother of Harold Arlen, who would help him break into the songwriting world. A prolific composer of tunes, Van Heusen collaborated with several lyricists, notably Johnny Burke and Sammy Cahn. Van Heusen and Burke moved to Hollywood in 1940 to focus on film work and won the Best Original Song Oscar for "Swinging on a Star," written for *Going My Way* (1944) starring Bing Crosby. It was the first of Van Heusen's ten Academy Award nominations. Van Heusen and Cahn wrote such standards as "All the Way," "High Hopes," and "Call Me Irresponsible." Van Heusen claims the inspiration for his pen name was a Van Heusen Shirts billboard that stood atop a building across the street from the Hotel Syracuse.

BILL VIOLA (1951–), video artist, is among the pioneers of independent video art and is recognized around the world as one of its most accomplished practitioners. Viola came to Syracuse University from Long Island in the late 1960s to study in the College of Visual and Performing Arts' experimental studios program, earning a bachelor of fine arts degree in 1973. His first job was working videotech for the Everson Museum in downtown Syracuse, and he then served as video artist in residence at WNET in New York City. His work has been presented at top contemporary art museums, including the Museum of Modern Art in Manhattan, the Getty in Los Angeles, and the National Gallery in London. An exhibition in Tokyo, *Bill Viola: Hatsu-Yume* (Bill Viola: First Dream, 2006) drew 340,000 visitors to the Mori Art Museum during a three-month engagement. Viola received an honorary Doctor of Fine Arts degree from Syracuse in 1995.

M. EMMETT WALSH (1935–), actor, was born in Ogdensburg (St. Lawrence County) and graduated from Clarkson University in Canton. Since his film debut in *Alice's Restaurant* (1969), Walsh has become a remarkably prolific screen actor, often appearing in as many as six movies per year. A rumpled figure, he is frequently cast as a corrupt cop, a shady ne'er-do-well, or a sleazy redneck. His films include *Blade Runner* (1982) with Harrison Ford; *Blood Simple* (1984); *Twilight* (1998), starring Paul Newman and Susan Sarandon; and *Snow Dogs* (2002). Critic Roger Ebert thought so much of Walsh's abilities that he created the "Walsh Rule," stating that no film in which he appears can be altogether bad.

LAWRENCE EDWARD WATKIN (1901–1981), fiction writer and screenwriter, hailed from Camden (Oneida County) and attended Syracuse University, where he earned a bachelor's degree in 1925 and a master's degree a year later. His novel *On Borrowed Time* 1937), an engrossing Irish fable of Death imprisoned in an apple tree, was the source for a 1939 feature film starring Lionel Barrymore and Cedric Hardwick. In the late 1940s Watkin turned to writing for the movies, which he did, prolifically, for Walt Disney Pictures. Watkin screenplays include *Treasure Island* (1950), *The Story of Robin Hood and His Merrie Men* (1952), *The Sword and the Rose* (1953), *Rob Roy, the Highland Rogue* (1954), *The Great Locomotive Chase* (1956), and *Darby O'Gill and the Little People* (1959).

JOHN VAN ALSTYN WEAVER (1893–1938), playwright and screenwriter, was born in Charlotte, North Carolina, and graduated from Hamilton College in 1914. *Love 'Em and 'Leave 'Em* (1926), starring Louise Brooks, was based on Weaver's stage play of the same name. Weaver's first screenplay was for *The Crowd* (1928), a silent feature directed by King Vidor. The following year he collaborated on a talkie, *The Wild Party* (written by Samuel Hopkins Adams under the pseudonym

85. Clara Bow, the "It" girl, was the star of the 1929 film *The Wild Party,* based on a screenplay by John Van Alstyn Weaver, from the novel *Unforbidden Fruit* by Samuel Hopkins Adams. Library of Congress, Prints & Photographs Division, LC-DIG-ggbain-38839.

Warner Fabian), starring Clara Bow and Fredric March, a precode picture about the out-of-control coeds. Weaver wrote just seven more screenplays, including an adaptation of Twain's *The Adventures of Tom Sawyer* (1930), before tuberculosis ended his life at age forty-four.

CHRIS WEDGE (1957–), director of animated films, was born in Binghamton and later attended Fayetteville–Manlius High School in suburban Syracuse, graduating in 1975. He went on to earn a bachelor of fine arts degree at SUNY Purchase in Westchester and a master's degree in computer graphics at Ohio State. Wedge, who won an Academy Award for his animated short "Bunny" (1998), got his start in Hollywood on the crew of *Tron* (1982). He has codirected two animated features with Carlos Saldanha, *Ice Age* (2002) and *Robots* (2005), and is producing a screen version of the Dr. Seuss children's book, *Horton Hears a Who.*

PETER WELLER (1947–), actor, earned a bachelor's degree in theater from the American Academy of Dramatic Arts and debuted on Broadway just months later in the New York Shakespeare Festival's production of *Sticks and Bones.* He has appeared in more than sixty film and television roles, ranging from the title role in the pop sci-fi *RoboCop* (1987) to the lead in *Naked Lunch* (1991), director David Cronenberg's remarkable adaptation of William Burroughs's surrealistic novel. Weller was already an accomplished actor and movie star when his interest in art history led him to enroll in graduate school at Syracuse University. He earned his M.A. degree in 2005.

EDWARD NOYES WESTCOTT (1846–1898), fiction writer and banker, was a Syracusan. He wrote *David Harum* (1898; see plate 22), a novel concerning a homespun banker who was also a horse trader. This was filmed twice, as a silent in 1915 with William Crane, and as a talkie in 1934 with Will Rogers. The story's setting, Homeville, was purportedly Homer, a village south of Syracuse in Cortland County.

FRANK WHALEY (1963–), actor and director, was born and raised in Syracuse and went to college at SUNY Albany. He has appeared in more than sixty film and television productions, including such features as *Ironweed* (1987, set in Albany), *Field of Dreams* (1989), *Born on the Fourth of July* (1989), *The Freshman* (1990), *Pulp Fiction* (1994), *World Trade Center* (2006), and *Aftermath* (2007). Whaley wrote and directed *New York City Serenade* (2007), a comedy starring Freddie Prinze, Jr., and Chris Klein.

LEOPOLD WHARTON (1870–1927) and THEODORE WHARTON (1875–1931), brothers, were producers and directors who established the Wharton Brothers Film Studios in 1914 on leased land in Ithaca, near what is now Stewart Park, on the southern shore of Cayuga Lake. The brothers had been to Ithaca in 1912 to make *Dear Old Girl of Mine*, a romance set on the Cornell University campus, for Essanay Studios. They wanted to start a studio of their own and saw possibilities for outdoor shooting in Ithaca's extraordinary topographical features: gorges, waterfalls, lakes, forests, and rock formations. They used these assets to great advantage. For example, in the making of *A Prince of India* (also known as *Kiss of Blood*, 1914), the Whartons purchased a trolley car and filmed it as it left the track on the Stewart Avenue bridge and plummeted 150 feet into the gorge below. Shot at midafternoon on a sunny August day, the spectacular stunt attracted most of the town, and not all the residents were prepared to understand the nuances of art and life involved in it. For the filming of *The Shanghai Man* (1914), the Whartons cast fifty Onondaga Indians to play a tribe of Aztecs in full costume. Brought by train from the Onondaga Nation reservation, just south of Syracuse, the actors were met at the Ithaca railway station by a special trolley that brought them to the studio. During this period, federal law prohibited the serving of alcoholic beverages to reservation Indians, and Ithaca police officers were on hand to enforce the federal ordinance. Onondaga tribal chief Harry Isaacs made his film debut at age seventy by performing a successful fifty-foot dive into Fall Creek Gorge, wearing full tribal regalia. In less than seven years, more than seventy silent screen titles were produced by the Wharton Brothers at their Ithaca studio.

86. Pearl White in a rural pose, with a pig on her lap, ca. 1916. Library of Congress, Prints & Photographs Division, LC-USZ62-57953.

E. B. WHITE (1899–1985), fiction writer and essayist, was born in White Plains (Westchester County) and attended Cornell University. After graduating in 1921, he served as a reporter for several New York City papers and joined the staff of *The New Yorker*, which he would be associated with for most of his career. White was a prolific writer, but Hollywood was interested primarily in his children's books. Three were made into animated features: *Charlotte's Web* (1973), with voices by Debbie Reynolds, Paul Lynde, and Agnes Moorehead; *Stuart Little* (1999), with Michael Fox speaking the title role; and *Trumpet of the Swan* (2001), with voices by Jason Alexander and Carol Burnett.

PEARL WHITE (1889–1938), star of the popular serial *The Exploits of Elaine* (1914–15), was criticized by some of the Ithaca town folk for her unconventional clothing (she often wore slacks), for smoking in public places, and for her unbridled use of profanity. She took particular delight in driving around Ithaca in her expensive convertible, earning as many speeding tickets as police could write. Members of Ithaca's "smart set," by contrast, were enamored of White and their other new neighbors. Cornell ornithology professor Louis Fuertes arranged for his toddler son, Louis, Jr., to take the title role in *Little Ned* (1913), and encouraged his wife, Summer, to appear in the film as well. Fuertes attended shooting sessions, learning camera techniques from the Wharton crew that he applied to his study of birds.

JOHN ALFRED WILLIAMS (1925–), journalist and fiction writer, was born in Jackson, Mississippi, and attended Syracuse University, graduating in 1950. He is cited by the *Dictionary of Literary Biography* as "arguably the finest Afro-American novelist of his generation, although he has been denied the full degree of support and acceptance some critics think his work deserves." During his versatile career, Williams has worked as a journalist for CBS News, *Ebony*, *Jet*, and *Newsweek*, and taught at several institutions, including Rutgers University, where he was named the Paul Robeson Professor of English in 1990. His 1961 novel *Night Song* was adapted for the screen as *Sweet Love, Bitter* (1967), starring Dick Gregory. *The Junior Bachelor Society* (1976), a Williams novel concerning a reunion of an African American social club, was the source for *The Sophisticated Gents* (1981), a made-for-television movie, with a script by Melvin Van Peebles.

VANESSA WILLIAMS (1963–), actress, from Millwood (Westchester County), studied theater at Syracuse University. While a student, she won the Miss Syracuse, Miss New York, and Miss America (1984) contests. Williams has a highly successful career as a recording artist who has sold more than six million albums; as a stage actress who has appeared on Broadway and at the Kennedy Center; and on television, where she has appeared on scores of series. Her feature film roles include *Soul Food* (1997), for which she received the NAACP's Image Award; *Shaft* (2000), *My Brother* (2006), and *And Then Came Love* (2007).

EMANUEL L. WOLF (1927–), producer, graduated from Syracuse University in 1950 and earned a master's degree there at the Maxwell School in 1952. He worked as president and chairman of the Board of Allied Artists Pictures Corporation and served as producer of such films as *Last Summer* (1969), *Friendly Persuasion* (1975), *The Betsy* (1978), and *Voyage to the Outer Planets and Beyond* (1986).

TOBIAS WOLFF (1945–), fiction writer and essayist, is a master of the memoir and the short story who taught in the creative writing program at Syracuse University from 1980 to 1997, before joining the faculty of Stanford University. He is best known for his 1989 memoir of childhood that was adapted into *This Boy's Life* (1993), starring Leonardo DiCaprio in one of his first major roles, with Robert DeNiro and Ellen Barkin.

LOUIS R. "WOLLY" WOLHEIM (1880–1931), actor, was born in New York City and attended Cornell University, where he played varsity football and graduated in 1907 with a degree in mathematics. While working as a part-time math instructor at Cornell in 1914, he was discovered by Lionel Barrymore, who helped him get uncredited roles in several films made at the Wharton Brothers's studio in Ithaca, including *The Warning* (1914) and *The Romance of Elaine* (1915). As Barrymore had surmised, Wolheim's nose, deformed by a football injury, made him a natural for parts as tough guys and thugs. Wolheim followed the movies to Hollywood and went on to appear in more than fifty films, including a stellar performance as Katczinsky in the Academy Award–winning *All Quiet on the Western Front* (1930). In contrast to his crude screen presence, he was an accomplished intellectual, a mathematician able to read in four languages. He died in 1931 from stomach cancer while working on what would have been his next film, *The Front Page* (1931).

ALEXANDER WOOLCOTT (1887–1943), playwright, critic, and actor, was a 1909 graduate of Hamilton College. A highly successful playwright, he collaborated on several shows with George S. Kauffman and was a newspaper columnist and critic as well. Woolcott's involvement with the movies was sporadic. He collaborated with Louis Bromfield on the story for the silent comedy *Bobbed Hair* (1925). His play, *The Dark Tower*, was put to film twice, first as *The Man with Two Faces* (1934) and then as *The Dark Tower* (1943). Woolcott also appeared in at least one feature film, *The Scoundrel* (1935), as well as several comic shorts. A member of the Algonquin Round Table, the famed group of literary raconteurs who met weekly at the Algonquin Hotel in Manhattan, Woolcott was portrayed on screen as a character in at least two films: *The Man Who Came to Dinner* (1942) and *Mrs. Parker and the Vicious Circle* (1994). Samuel Hopkins Adams, a fellow Hamilton College alumnus, wrote Woolcott's biography.

GRACE MILLER WRIGHT (1868–1965), fiction writer, lived in Ithaca. She wrote *Tess of the Storm Country* (1909; see plate 23), a novel about the daughter of a poor Cayuga Lake fisherman living in a squatter's village, who becomes enamored of an aristocratic Cornell University student. It was brought to the screen under the same title four times. Mary Pickford and Frederick Graves starred in two silent versions (1914 and 1922), Janet Gaynor and Charles Farrell in the 1932 talkie, and Diana Baker and Jack Ging in a 1960 Cinemascope remake that drops Cayuga Lake in favor of a Pennsylvania village. Other Wright novels were sources for *Secret of Storm Country* (1918), a Norma Talmadge vehicle made in Ithaca, and *From the Valley of the Living* (1915).

HAROLD BELL WRIGHT (1872–1944), fiction writer, was born in Rome (Oneida County) and lived in Sennett, a township on the northern fringe of the Finger Lakes, east of Auburn. Three Wright stories were each made as silents and remade as talkies: *The Shepherd of the Hills* (1928, 1941; see plate 24), *Mine with the Iron Door* (1924, 1936), and *When a Man's a Man* (1924, 1935). *The Winning of Barbara Worth* (1926), with Ronald Colman and Vilma Banky, is based on a Wright story as well.

APPENDIXES

 Appendix A

Theater Organs

Table A.1.

Theater Organs in Central New York

Theater	Organ Manufacturer	Yr. Installed	Opus Number	Manuals	Ranks
Solvay					
Allen	Link	1927			
Syracuse					
Avon	Wicks	1920	00306		
Avon	Marr and Colton	1926		3	8
Bill Purdy	Wurlitzer 235 NP	1927			
Brighton	Marr and Colton	1928		3	10
Cameo	Marr and Colton	1926			
Civic	Marr and Colton	1927		3	10
Crescent	Marr and Colton	1920		1	2
Eckel	Marr and Colton	1916			2
Eckel	Wurlitzer F3M	1927	01595	3	8
Elmwood	Link	1927		2	4
Empire	Marr and Colton	1925		3	10
Empire State Museum	Wurlitzer B	1964	01143	3	11
Happy Hour	Marr and Colton	1920			
Harvard	Wurlitzer B	1926	01282	2	4
RKO Keith's	Wurlitzer B 235	1925	01143	3	11
Lowe's State	Wurlitzer B SP4M	1928	01825	4	20
Palace	Wurlitzer B 109	1925	01159	2	3
Paramount	Wurlitzer E	1927	01761	2	7
Regent	Austin	1914	00485	2	12
Regent	Marr and Colton	1926		3	13
Riviera	Wurlitzer 175x	1928	01977	2	4
Rivoli	Marr and Colton	1922		2	6
Savoy	Marr and Colton	1920		2	8
Strand	Austin	1915	00545	2	11

(Continued on next page)

Theater	Organ Manufacturer	Yr. Installed	Opus Number	Manuals	Ranks
Strand	Wurlitzer 235 NP	1927	01662	3	11
System	Marr and Colton	1924			
Temple	Marr and Colton			2	7
Auburn					
Jefferson	Link	1926		2	
Palace	Wurlitzer EX	1927	01623		
Strand	Marr and Colton	1925		3	11
Universal	Marr and Colton	1921		2	
Oneida					
Madison	Marr and Colton	1927		3	9
Rome					
Capitol	Moller	1928	5371	3	7
Carroll	Marr and Colton	1921	3		
Utica					
Avon	F3M	1928	1874		
Avon	Wurlitzer 3	1915	71		
Deluxe (Oneida Sq.)	Buhl	1924			
Gaiety (Utica)	Wurlitzer 165x	1928	1966		
James	Link	1927		2	
Majestic	Marr and Colton				
Olympic	Marr and Colton	1926		2	
Park	Moller 1	1923	3123	3	10
Proctor H.S.	Wurlitzer 240				
Robbins (Lyric)	Marr and Colton	1921		2	
Schubert (Colonial)	Buhl	1915		3	
Slotnick	Link	1926			
South (Orpheum)	Marr and Colton	1917			
State	Buhl	1915		3	
Strand (Stanley)	Wurlitzer 240 SP	1928	1886		
Uptown	Marr and Colton	1927		3	11

Table A.2.

Organ Manufacturing Companies

Company	Location	Yrs. of Operation
Austin	Hartford, Connecticut	1993–
Beman	Binghamton, New York	1919–33
Buhl	Utica, New York	1905–50
Carey	Troy, New York	1965–
Delaware	Tonawanda, New York	(closed)
Estey	Brattleboro, Vermont	1901–59
Frazee	Everett, Massachusetts	1910–69
Gottfreed	Erie, Pennsylvania	1994–50
Hall	West Haven, Connecticut	1898–45
Holtkamp	Cleveland, Ohio	1951–
Hope-Jones	Elmira, New York	1907–10
Kimball	Chicago, Illinois	1957–42
Kohl	Rochester, New York	?–1954
Link	Binghamton, New York	1916–32
Marr and Colton	Warsaw, New York	1915–32
Midmer-Losh	Merrick, New York	1924–
Moller	Hagerstown, Maryland	1975–92
Roosevelt	New York, New York	1972–93
Schlicker	Buffalo, New York	1930–
Skinner	Boston, Massachusetts	1932–71
Wurlitzer	N. Tonawanda, New York	1904–

Appendix B

Drive-in Theaters in Central New York, Past and Present

The following tables list drive-in movie theaters in central New York including operating drive-ins (Table B.1.) and those that have closed (Table B.2.). The latter list of central New York's fallen screens is incomplete. Other closed drive-ins in the region include Carroll's Marcy, the Cobleskill, East (Auburn), Homer, Kallet's Marcy, Northside (Watertown), Norwich, Parkway (Canandaigua), Seneca (Geneva), and Skyway (Sodus), but only the sketchiest information about them is available.

Table B.1.

Operating Drive-in Movie Theaters in Central New York, 2007

Theater	Location	Cars	Yr. Built
Bay (twin since 1999)	Alexandria Bay	300/200	1968
Black River	Watertown	500	1951
Elmira (twin since 2000)	Elmira	400/400	1949
El Rancho	Palatine Bridge	300	1952
Finger Lakes	Auburn	300	1950
Midway	Minetto	600	1950
Papa's Place (formerly Bath)	Bath	(unknown)	1952
Unadilla	Unadilla	(unknown)	(unknown)
Valley Brook	Lyons Falls	150	(unknown)
West Rome (twin since 1985)	Rome	400/300	1951

Table B.2.

Sampling of Closed Drive-in Movie Theaters in Central New York

Theater	Location	Cars	Yrs. of Operation
Airport	Cortland	400	1970–?
Airport	Johnson City	1000	1961–93
C Way (formerly Free's)	Ogdensburg	(unknown)	1951–?
Del Sego	Oneonta	400	1954–81
DeWitt	DeWitt	800	1950–84
Dryden	Dryden	800	(unknown)
Front	Nimmonsburg	800	1950–58
Kallet	Camillus	800	1946–60
Kallet's (formerly WGAT)	New Hartford		1950–75
Lakes Car	Ithaca	(unknown)	(unknown)
Lakeshore	Liverpool	600	1957–89
New Hartford	New Hartford	750	1950–82
North	Cicero	700	1949–86
Northside	Watertown	(unknown)	1955–85
Salina	Nedrow	1200	1954–80
Skyler Twin	Utica	900	1968–83
Starlite	Watertown	275	1950–85
The V	Vestal	(unknown)	1961–87

 Appendix C

The Schine and Kallet Theater Chains in Central New York

The Schine circuit in New York State was administered as two divisional units, headquartered in Albany and Buffalo. The theaters listed in the first table, alphabetically by town, were controlled by the Schines during all or part of their existence. The Kallets ran their chain from their office on Madison Street in Oneida. The theaters listed in the second table, alphabetically by town, were controlled by the Kallets during all or part of their existence. For both chains, each theater's name is followed by its years of operation as a movie venue and its seating capacity, where known. Name changes and changes in seating capacity are noted in parentheses.

Table C.1.

Schine Theaters in New York State

Division/Town/Theater	Yrs. of Operation	Seating
ALBANY DIVISION		
Amsterdam		
Lyceum (Mohawk)	1919–66	
Regent	1931–57	1150
Rialto	1926–66	140
Strand	1926–55	1200
Ballston Spa		
Capitol	1931–48	600
Carthage		
Strand	1929–66	650
Dolgeville		
Strand	1925–35	450
Glens Falls		
Empire	1931–45	982
Rialto	1931–66	1291
Gloversville		
Family (Kassen Opera House)	1920–?	
Glove	1920–72	1200

(Continued on next page)

Division/Town/Theater	Yrs. of Operation	Seating
Hippodrome	1917–61	1200
Granville		
Ritz	1926–66	500
Hamilton		
State	1938–66	595
Herkimer		
Liberty	1922–69	1081
Richmond	1922–57	800
Hudson Falls		
Strand	1931–66	693
Ilion		
Capitol	1930–66	900
Opera House	1926–?	
Temple	1925–55	300
Little Falls		
Gem	1929–?	
Hippodrome	1922–55	800
Rialto	1931–66	1200
Lowville		
Bijou	1926–29	400
Malone		
Grand	1926–29	800
Malone	1922–66	1227
Plaza	1926–45	600
Massena		
Massena	1931–66	1065
Opera House	1929–?	
Rialto	1926–45	800
Strand	1929–?	500
Mechanicville		
State	1931–55	1129
Norwich		
Colonia	1925–66	800
Strand	1926–29	900
Ogdensburg		
Pontiac	1939–55	1092
Star	1926–29	200
Strand	1926–66	1400
Oneonta		
Oneonta	1922–66	450

(Continued on next page)

Division/Town/Theater	Yrs. of Operation	Seating
Palace	1931–57	1045
Strand	1922–29	
Saranac Lake		
Pontiac	1929–66	1046
Tupper Lake		
State	1931–57	500
Watertown		
Avon	1920–66	702
Olympic	1920–66	100
Palace	1920–51	300
Whitehall		
Capitol	1931–55	800
BUFFALO DIVISION		
Auburn		
Auburn	1938–78	1800
Capitol	1927–51	1100
Jefferson	1908–57	1300
Palace	1927–64	1075
Strand	1925–53	1725
Bath		
Babcock	1935–66	693
Buffalo		
Granada	1929–66	1746
Riverside	1929–57	1600
Canandaigua		
Lake	1942–51	374
Playhouse	1922–66	1131
Corning		
Fox	1931–66	1358
Palace	1931–51	385
Regent	1929–?	
State	1929–51	926
Cortland		
State	1931–66	1702
Temple	1929–53	1800
East Rochester		
Rialto	1926–55	1020
Elmira		
Capitol	1929–48	1500
Colonial	1929–48	685

(Continued on next page)

Division/Town/Theater	Yrs. of Operation	Seating
Fairport		
Temple	1929–48	886
Geneva		
Geneva	1929–66	1862
Regent	1925–57	1862
(Smith Opera House)	1994–29	
Temple	1925–57	1000
Lockport		
Hi-Art	1925–56	700
Palace	1926–66	1750
Rialto	1926–57	1400
Temple	1929	
Newark		
Capitol	1925–66	1246
Crescent	1926–29	700
Oswego		
Capitol	1924–57	700
Oswego	1942–66	1800
State	1929–31	713
Strand	1924–57	895
Penn Yan		
Elmwood	1921–70	700
Perry		
Auditorium	1926–55	620
Rochester		
Cameo	1931–57	1200
Dixie	1931–48	860
Grand	1925–57	800
Lake	1931–57	700
Liberty	1929–55	450
Madison	1931–57	1200
Monroe	1931–66	1197
Riviera	1929–66	1057
State	1929–57	923
Webster	1929–45	900
West End	1931–57	860
Salamanca		
Andrews	1922–51	807
Seneca	1926–66	1272
Seneca Falls		
Strand	1928–57	987

(Continued on next page)

Division/Town/Theater	Yrs. of Operation	Seating
Syracuse		
Eckel	1914–78	1407
Palace	1926–present	830
Paramount	1928–67	1500
RKO Keith's	1920–67	2514
Strand	1914–59	1600
Utica		
DeLux	1916–35	1200

87. The James Theatre was a Kallet theater in Utica, New York. Courtesy of Robert Kallet, Joe Pfeiffer, and Bob Card.

88. The Uptown Theatre was another Utica theater belonging to the Kallet chain. Courtesy of Robert Kallet, Joe Pfeiffer, and Bob Card.

Table C.2.

The Kallet Theater Chain in New York State

Town/Theater	Yrs. of Operation	Seating
Avon		
Park	1926–55	400
Baldwinsville		
Palace	1934–65	700
Bolton Landing		
Rex (Kallet)	1935–64	400
Boonville		
Avon	1925–38	322
Franjo	1931–63	688
Brockport		
Strand	1931–69	500
Camden		
Arcade	1925–35	450
Camillus		
Camillus Drive-In	1945–59	800 cars
Canajoharie		
Strand	1931–55	576
Canastota		
Avon	1912–67	500
Deposit		
Empire	1935	400
State	1937–57	500
DeWitt		
Shoppingtown	1955–68	1004
Ellenville		
Norbury	1942–55	613
Fleischmanns		
Whipple	1939–45	600
Fulton		
Avon (Happy Hour)	1928–45	800
State (Quirk)	1913–57	1200
Geneseo		
Palace	1931–55	403
Riviera	1939–55	500
Ilion		
Capitol	1926–66	900 (1250)

(Continued on next page)

Town/Theater	Yrs. of Operation	Seating
Temple	1925–29	300
Le Roy		
Le Roy	1935–60	336 (640)
Loch Sheldrake		
Strand	1939–69	517
Long Lake		
Strand	1938–56	400
Marcy		
Kallet Drive-in		
Margaretville		
Galli-Curci	1922–57	500
Mineville		
Rivoli	1939–51	300
New Hartford		
Olympic Drive-in		
Oneida		
Kallet	1938 69	1200
Madison	1922–53	960
Regent	1925–31	600
Port Henry		
Essex	1942–57	560
Pulaski		
Kallet	1939–69	558 (680)
Temple	1925–35	565
Rome		
Capitol	1928–77	1500 (2096)
Family (Lyric)	1907–54	
Star	1913–28	500
Strand (Carroll)	1911–65	1100
South Fallsburg		
Rivoli	1925–69	700
Syracuse		
Avon	1928–53	700
Dreamland	1910	
Genesee	1950–97	1077
Harvard (Westcott)	1925–2007	746
Regent	1914–57	1000
Riviera	1928–72	860
Ticonderoga		
State	1939–69	444

(Continued on next page)

Town/Theater	Yrs. of Operation	Seating
Utica		
James	1927–53	1200
Kallet Drive-in		
Kallet's Drive-in		
Liberty		500
Lincoln	1927–53	800
Oneida Square	1935–55	1200
Orpheum	1917–53	1100
Warsaw		
Farman	1931–66	1000
O-At-Ka	1926–53	444
Woodburne		
Center	1939–66	670
Woodridge		
Lyceum	1931–69	300

Appendix D

Central New York Movie Theaters

Theaters are listed alphabetically and numerically by street address for the city of Syracuse, and alphabetically by town for other locations in Central New York, including street addresses when known. Each theater listing contains the following information, if known: changes in name or function (indicated by the symbol →); number of seats; and years of operation. In cases where the year of a name change is not known, the new name is indicated with *aka* ("also known as"). Multiscreen mall cinemas, whether operating or closed, are not included in this listing of theaters.

Table D.1.

Theaters in the City of Syracuse by Location

Street/No.	Theater Name	Seating	Yrs. of Operation
Burnet Avenue			
1601	Everybody's	450	1922–31
Burnet Park Drive			
(formerly S. Milton Ave.)			
109	Park	400	1913–53
113	Kernan	400	1920–31
Butternut Street			
718	Butternut → Carl Lang		1915–18
909	Acme	800	1915–53
Clinton Street E.			
119	Dunfee's Comedy		1898
119	Lyceum	1300	1903–4
Crouse Avenue S.			
112	The Crouse		1913–15
Division Street E.			
701	Poplar	275	1922–25

Theaters whose names are preceded by an asterisk (*) are currently in operation.

(Continued on next page)

Street/No.	*Theater Name*	*Seating*	*Yrs. of Operation*
Eagle Street			
511	Crystal		
942	Eagle		
Fayette Street E.			
216	Eckel	1600	1914–78
704	Little		1921–25
1205	Gilfoil Duncan		1913–15
Fayette Street W.			
213	Novelty	550	1909–53
218	Rivoli	1250	1922–53
243	Bijou		1909
	→ Lyceum		1912–15
Franklin Street S.			
319	Palace		1909–15
Geddes Street N.			
203	The Geddes		
	→ Pastime	300	1913–28
Geddes Street S.			
	Robert Burns		1914
	→ Model		
637	Castle	400	1912–15
656	Cameo	900	1927–60
1103	Langan aka Tony's Opera	360	1913–27
Genesee Street E.			
136	Larned		1909–14
207	Bastable	1045	1893–1923
312	Grand Opera House	2000	1886–1923
345	Gayety		1913
820	Regent	1000	1914–57
823	Irving		1913–14
Genesee Street W.			
2100	Genesee	1077	1950–97
1205	Pictureland		1913–16
Gifford Street			
508	Royal aka Thomas Haley		1913
Grant Boulevard			
(formerly Manlius Street)			
1865	Schiller Park	750	1914–42
	→ Grant		1946–53
Hawley Avenue			
433	East Side Family		1912

(Continued on next page)

Street/No.	Theater Name	Seating	Yrs. of Operation
	→ Colonial		1920–21
443	Avon	700	1928–53
615	Lincoln	300	1914–26
Irving Avenue			
	Varsity	500	1914–25
James Street			
275	Alhambra	5000	1884–54
2384	*Palace	700	1924–present
2811	Melva	474	1923–29
2819	James	400	1937–45
Kirkpatrick Street			
710	Globe	550	1922–60
	→ Actorgraph		1914–16
Lowell Avenue N.			
	Celtic		1913–18
Midland Avenue			
525	Elite		1915
	→ Victoria		1916–18
2215	Midland	1809	1913–22
Montgomery Street			
822	Starland		1913–25
Onondaga Street E.			
218	Tivoli		1914
Onondaga Street W.			
109	Dreamland	120	1910–16
Oswego Street			
605	Capitol	600	1922–37
	→ Mayfair		1937–40
Park Avenue			
744	Liberty	500	1918–48
Park Street			
327	Pictorial		1914
	→ Hyperion		1920
	→ Highland		1921–25
1003	Park	300	1912
	→ Rialto		1921–22
Salina Street N.			
224	Happy Hour	500	1912
	→ Swan		1926–33
	→ Midtown		
245	Majestic		1911

(Continued on next page)

Street/No.	Theater Name	Seating	Yrs. of Operation
	→ Morgan's		
262	Rex aka Oswald		1914–16
547	Star	312	1909–39
621	Turn Hall	600	1912–54
626	Luna aka Princess		1910–14
839	Studio		1914–15
1636	Crystal		1912
Salina Street S.			
328	May's		1907
358	Hippodrome		1908–23
362	Loew's State	2900	1928–68
	→*Landmark Theater		?–present
	(perf. arts cntr.)		
408	RKO Keith's	2548	1920–67
424	Paramount aka Temple		1914–67
429	Palace		1914
451	Crescent	1000	1909–28
468	Empire	1772	1911
	→ Dewitt		1932
	→ Astor		1949–53
473	Majestic		1908–9
476	Antique		1909
501	Loew's Strand	1600	1914–59
544	Salina		1914
572	Civic aka Syracuse, Top,		
	→ Ritz, System	1100	1914–76
1010	Mecca		1912–14
1140	Arena	6000	1911–39
2203	Arcadia aka Colvin	550	1914–31
2616	Brighton	1750	1929–60
2622	Lafayette		1912–16
2706	Plaza	400	1913–31
3120	Riviera	1100	1929–72
Seneca Turnpike W.			
	Valley	2000	1899–1916
Seymour Street			
159	Seymour	500	1913–27
South Avenue			
237	Franklin	535	1914–78
318	Starlite		1915–16
1828	Elmwood	650	1914–59

(Continued on next page)

Street/No.	Theater Name	Seating	Yrs. of Operation
State Street S.			
1234	Colonial		1913–15
Townsend Street S.			
(formerly Grape Street)			
511	Crystal		1911
942	Eagle		1914–18
1332[a]	Alcazar	400	1913–53
Warren Street S.			
109	Standard		
	→ Rialto		1913–23
316	Savoy	1000	1909–31
Water Street W.			
110	Wieting Opera House	2210	1897–1932
Westcott Street			
475	Playhouse		1914–17
524	Harvard	746	1926
	→ Studio		1968
	→ Westcott Cinema		1971–2007
Willis Avenue			
1021	Kernan	300	1913–36
Wolf Street			
258	Lyric	350	1912–45

[a]This location's address is now 376 Oakwood Drive.

89. The Idle Hour, at 300 West Dominick Street in Rome, New York, was an early nickelodeon theater. Courtesy of the Rome Historical Society.

90. The Franjo Theatre in Boonville, New York. Photograph by James Fynmore. Courtesy of Robert Kallet and Joe Pfeiffer.

Table D.2.

Other Central and Upstate New York Movie Theaters by Town

Town/Location	Theater Name	Street	Seating	Yrs. of Operation
Adams				
	Masonic Hall		475	1927–29
	Opera House		207	1925–35
	State		325	1939–51
Alexandria Bay				
	Bay	S. Bethune St.	550	1940–55
	Weller's			1930–35
Auburn				
	Auburn Schine	16 South St.	1800	1938–78
	Burtis Auditorium		3200	1904
	Burtis Grand	20 North St.	600	1889–1913
	→ Capitol		645	1927–51
	Happyland	7 Green St.		1910
	→ Cornell			1912
	Hippodrome	10 South St.		1912–13
	Jefferson	61 State St.	1300	1908–52
	→ Morgan			1913
	→ Universal			1921
	Motion World	1331–32 Genesee St.	1200	1909
	→ Universal			1914–21
	Palace	60 Genesee St.	1075	1927–72
	Regent	7 Green St.		1919–2007
	Strand	35 Water St.	1725	1924–53
Aurora				
	Aurora		300	1926–29
	Morgan Opera House	1899 Main St.	200	1899–1942
Bainbridge				
	Avon	13 N. Main St.	275	1945–51
	Lyric			1925–35
	Town Hall		250	1931–45
Baldwinsville				
	Howard's Opera House		450	1881
	→ Grange			1914–21
	Orpheum		500	1911–35
	Palace	39 Oswego St.	800	1934–65
	→ Paramount			1933
	→ Varicty			1928
	Wonderland Nickel	15 Syracuse St.		1909

Theaters whose names are preceded by an asterisk (*) are currently in operation.

(Continued on next page)

91. The Quirk Theatre in Fulton, New York, was the predecessor to the State Theatre, and stood on East 1st Street. Courtesy of the Friends of History in Fulton, N.Y. Inc.

Town/Location	Theater Name	Street	Seating	Yrs. of Operation
Binghamton				
	Art	1204 Vestal Ave.	277	1935–2004
	Avon		450	1935
	Binghamton	236 Washington St.	1747	1925–51
	→ Capri			1951
	→*Forum Perf. Arts Cntr.			1975–present
	Cameo	234 Robinson St.	800	1929–2002
	Capital	60 Exchange St.	2284	1931
	City Lines		400	1925–35
	Crest	244 Main St.	804	
	Grand	1204 Vestal Ave.	500	1925
	Happy Hour		600	1925–39
	Hider		400	1925–31
	Jarvis	169 Main St.	600	
	Lyric	140 Water St.	906	1925
	→ Laurel		500	1925–35
	Peoples		650	1925–35

(*Continued on next page*)

Town/Location	Theater Name	Street	Seating	Yrs. of Operation
	Regus	316 Chenango St.	800	1925–51
	Ritz	50 Clinton St.	550	1931–91
	Riviera	33 Chenango St.	1695	1931–73
	Star	51 Chenango St.	800	1925–51
	Stone Opera House		1500	1925–31
	Strand	27 Chenango St.	1472	1925–55
	Suburban	244 Main	1015	1929–55
	Sun	70 Glenwood Ave.	600	1929–55
	Symphony	109 Chenango St.	700	1925–51
Boonville				
	Avon		372	1925–38
	Bijou		400	1929
	Comstock's Opera House			1929
	Franjo	18 Schuyler St.	688	1931–63
Bridgeport				
	Brown's Opera House		250	1927–31
Camden				
	Arcade		450	1925–35
	Park		500	1929
	→ Smalleys			1939–57
Camillus				
	Kallet Drive-in		800 cars	1945–59
	Star		300	1928–31
Canandaigua				
	Empire			1920
	Family			1920
	Lake	Main St	374	1942–51
	Liberty		1000	1920–26
	Playhouse	Chapin St.	1131	1922–66
	Temple		600	1924–26
Canastota				
	Sherwood	N. Peterboro St.		1912
	→ Avon		500	1914–67
Candor				
	Candor		250	1925–29
Cape Vincent				
	Strand	Main Street	214	1940–55
Carthage				
	Hippodrome		700	1926

(Continued on next page)

Town/Location	Theater Name	Street	Seating	Yrs. of Operation
	→ Strand	228 State St.	1265	1926–66
	Temple		600	1931–34
Cayuga				
	Mansfield Hall			
Cazenovia				
	Town Hall	10 Lincklaen St.	409	1928–74
Central Square				
	Masonic Temple		300	1928
Chadwicks				
	Men's Club		700	1926–29
	Standard Hall		400	1925–39
	Willowale			1925
Cherry Valley				
	Cherry Valley		300	1935–51
	Star			1924–31
Chittenango				
	Carl Opera House			1910→
	Laning			1929–31
	Delphia	Seneca St.	410	1929–58
	Union Hall			1914–21
Cincinnatus				
	Alhambra		250	1929–34
	Avon		250	1925–29
Cleveland				
	Home		200	1931
	Novelty		200	1926–29
Clifton Springs				
	Palace	10 Crane St.	400	1927–57
Clinton				
	Clinton Heights		200	?–1945
	Garvey's		300	1925
	Grange aka Clinton	2 Fountain St.	300	1929–55
Clyde				
	Playhouse		300	1927–55
	Utopian		300	1925–29
Constableville				
	Star		250	1925–29
Cobleskill				
	*Park	1 Park Plaza	650	c. 1930–present

(Continued on next page)

Town/Location	Theater Name	Street	Seating	Yrs. of Operation
Corning				
	Fox	Erie Ave.	1358	1931–55
	Liberty		1200	1925
	Palace	W. Market St.	450	1945–51
	Plaza	3 W. Williams St.	460	1925–55
	Princess		650	1925–29
	Regent		450	1925–31
	State	Pine St.	926	1925–51
Cooperstown				
	Smalley's	137 Main St.	702	1931–55
Cortland				
	Cortland		1200	1925–39
	Novelty		600	1926
	Paramount		500	1925–31
	→ Schine		1653	1931–35
	State	27 N. Main St.	1753	1931–57
	Temple	9 Groton Ave.	656	1926–53
Cranberry Lake				
	Park		200	1927–29
Croghan				
	Wonderland		300	1925–29
Deansboro				
	Men's Club		200	1928
Delhi				
	Smalley's aka → Opera House	Kingston St.	450	1924–55
Deposit				
	Empire		400	1935
	Scott's Oquaga Lake → House Theater aka → Scott's Casino		500	1951–55
	*State Theater	148 Front Street	514	1937–present (intermittent)
DeRuyter				
	Opera House		250	1925
	Union Hall		275	1928–31
DeWitt				
	Shoppingtown	3649 Erie Boulevard E.		1955–68

(Continued on next page)

Town/Location	Theater Name	Street	Seating	Yrs. of Operation
Dolgeville				
	Smalley's		677	1939–55
	Strand			1924–31
Downsville				
	Colchester Theater aka			1924–55
	→ Opera House			
Dryden				
	Opera House	.	500	1893–63
	Totman		300	1928
Dundee				
	Beckman		400	1926–35
	Strand aka Dundee	Water St.	250	1928–51
Eagle Bay				
	Brown's			1925–28
Earlville				
	Douglas		500	1926–31
	Earlville		265	1939–53
	Opera House		300	1925
East Syracuse				
	Roxie	Manlius St.	700	1929–39
	→ East Theater			1939–61
	Steele		350	1925–31
Edmeston				
	Edmeston	Main St.	300	1928–57
	Picture			1924–?
Elmira				
	Amuzu			1925–29
	Capitol	114 State St.	900	1929–55
	Colonial	123 N. Main St.	1183	1925–80
	Elmira	111 College Ave.		1929–51
	Keeney's	219 State St.	2362	1925
	→*Samuel Clemens Cntr. for Perf. Arts			1977–present
	Lyceum	Lake St.		c. 1900–29
	Majestic	101 W. Market St.	1450	1929–31
	Mozart	E. Market St. at Lake St.		
	Regent	100 N. Water St.	850	1924–55
	Star	Pennsylvania Ave.		
	Strand	313 E. Market St.	1000	1924–51

(Continued on next page)

Town/Location	Theater Name	Street	Seating	Yrs. of Operation
Elmira Heights				
	*Heights	210 E.14th St.	720	1949–present
Endicott				
	*Cinema Saver Five	19 Madison Ave.		
	Elvin	117 W. Main St.	788	1928–
	Lyric	102 Washington Ave.	666	1917
	→ Towne			c. 1960–93
	→*Endicott Perf. Arts Cntr.			1998–present
	State	108 E. Main St.	605	1938–
	Strand	123 Washington Ave.	782	1925–
Fair Haven				
	Adelphi		250	1926
	Lakeside		275	1924–29
Fayetteville				
	Groveland	Mill St.	400	1914–29
Frankfort				
	Hollywood		400	1938–52
	Loomis		600	1928–
Fulton				
	Community		900	1925
	Happy Hour	119 Cayuga St.	800	1928
	→ Avon			1945
	Quirk	103 S. First St.	1000	1913
	→ State			1937–57
Geneseo				
	Grand		400	1924–26
	Palace		403	1931–55
	Rex		600	1924–35
	Riviera		500	1935–55
Geneva				
	Geneva	82 Seneca St.	1862	1931–55
	Park		400	1938–43
	Regent	499 Exchange St.	1000	1925–56
	Smith Opera House		982	1894
	Temple		530	1925–43
Genoa				
	Gem		250	1925–35
Georgetown				
	Georgetown			1929
	Town Hall		300	1927–29

(Continued on next page)

Town/Location	Theater Name	Street	Seating	Yrs. of Operation
Gloversville				
	Darling			
	Family	26–38 N. Main St.		
	Glove	42 N. Main St.	1200	1914–76
	Hippodrome	34 E. Fulton St.	1200	1914–52
Groton				
	Corona		500	1939–55
	Groton		690	1925–31
Hamilton				
	*Hamilton Movie House	7 Lebanon St.	370	1895–present
	*Palace (perf. arts cntr.)	19 Utica St.		?–present
	Smalley's	7 Lebanon St.	600	1924
	→ *State			1938–present
Hancock				
	*Capra Cinema	533 W. Front St.		?–present
Harrisville				
	Capital		160	1935
	Diane		225	1939
	→ Royal			1942–51
	Empress			1925
	Lyric		400	1924
	→ Castle			1931
Henderson				
	Henderson Par. House		200	1929
Henderson Harbor				
	Associated Island Corporation		250	1929
Herkimer				
	Hippodrome			
	Liberty	156 N. Main St.	900	1922–69
	Richmond		800	1922–35
Homer				
	Capitol		496	1945–55
	Community		500	1925–31
Ilion				
	Big Ben	First St.	175	1922–29
	Capitol		1250	1926–66
	Opera House			

(Continued on next page)

Town/Location	Theater Name	Street	Seating	Yrs. of Operation
	Temple		300	1925–29
Indian Lake				
	Lake	Main St.	300	1945–57
	Pelon's		150	1926–29
Inlet				
	Gaiety		250	1929–55
Interlaken				
	Lakes		230	1931–51
	Melville		400	1925–29
Ithaca				
	Clinton Hall	116–18 N. Cayuga St.		
	Crescent	215 N. Aurora St.	1300	1924–
	Happy Hour	115 N. Tioga St.	700	1924–
	Lyceum	113 S. Cayuga St.	1200	1893
	Ryan's Ithaca	314 W. State St.	600	
	Star	118–20 E. Seneca	1200	
	State	117 W. State	1626	1928–85
	→*(perf. arts cntr.)			2001–present
	Strand	310 E. State St.	1650	1917–76
	Temple	114 E. Seneca St.	850	1929–57
Johnson City				
	EnJoy	32 Willow St. aka Goodwill[a]	1111	1928–60; 1974–77
Johnstown				
	Smalley's	20 N. Market St.	1055	1928–55
Jordan				
	Hippodrome		200	1925–31
Lake Placid				
	Club		1000	1926–31
	Happy Hour			1924–27
	*Palace	26 Main St.	921	1926–present
Lakeside				
	Lakeside Park Rustic		2500	1999–2006
Leonardsville				
	Crescent		300	1926–35
Little Falls				
	Gateway			1925
	Hippodrome	S. William St.	800	1922–55
	Lintonian			1922–25

[a]Under redevelopment as Goodwill Theater Performing Arts Center

(Continued on next page)

Town/Location	Theater Name	Street	Seating	Yrs. of Operation
	Rialto	13 N. Ann St.	1100	1929–55
Liverpool				
	Lakeshore → Liverpool	303 First St.	215	1915; 1924–31
Long Lake				
	Community		200	1931–35
	Strand		400	1938–56
Loon Lake				
	Casino		250	1926–35
Lowville				
	Avalon		587	1931–57
	Bijou		400	1924–29
	Opera House		400	1927–35
	*Town Hall		750	1951–present
Luzerne				
	Burts		308	
Lyons				
	*Ohmann	65 William St.	650	1925–55; ?–present
Lyons Falls				
	McAlpine		350	1925–29
Manchester				
	Pastime		250	1925–34
Manlius				
	*Manlius Art Cinema	135 E. Seneca St.	250	1918–present
	Marion aka Fowler Hall			1914
	Smith Hall			1940
	St. John's Academy		500	1925
	Star	Mill St. and Seneca St.		1913
	Strand aka Seville, Lincoln, Colonial, Manlius	135 E. Seneca St.	280	1918–present
Marathon				
	Library Opera House		400	1925–38
	Park	214 E. Main St.	367	1945–57
Marcellus				
	Parson's Hall aka Strand	Main St.	300	1929–57
Margaretville				
	Galli-Curci		500	1922–57

(Continued on next page)

Town/Location	Theater Name	Street	Seating	Yrs. of Operation
Mattydale				
	*Hollywood	2221 Brewerton Road	600	1925–present
Mexico				
	Jasmine		300	1939
	Mexico	Main St.	400	1945–55
	Star		300	1925–29
	Washington		175	1931
Minetto				
	Community		400	1929
Mineville				
	Rivoli		300	1939–57
Mohawk				
	Bates Opera House		518	1928–35
	Strand			1925
	Smalley's			
Montour Falls				
	Opera House		250	1925–29
Moravia				
	Colonial		250	1937–55
	Opera House		350	1925–29
	Moravia			1931
Morrisville				
	Madison Hall	Rt. 20	300	1929
	Morrisville aka Morriss		200	1925–57
Newark				
	Capitol		1310	1925–66
	Crescent		700	1925–29
	Granite Opera House		600	1926–29
	Newark		700	1931–84
	*Newark Showplace	101 S. Main St.		?–present
Newark Valley				
	Opera House		400	1926–35
New Berlin				
	Dakin Hall		430	1934–55
	Opera House	N. Main St.		1926
Newcomb				
	Community		150	1924–29
New Hartford				
	New Hartford	12 Genesee St.	482	1951–53
	Players		300	1942–44

(Continued on next page)

Town/Location	Theater Name	Street	Seating	Yrs. of Operation
Newport				
	Star		500	1925–29
Newton Falls				
	Community		120	1939–51
	Newton Falls		300	1929–35
	Victory			1925
New York Mills				
	Atlas		400	1925
North Rose				
	Palace		250	1925–29
North Syracuse				
	Community		350	1925–31
Norwich				
	*Colonia	35 Broad St.	1000	1925;
				?–present
	Smalley's		854	1931–57
	Strand		900	1925–29
Odessa				
	Opera House		300	1925–29
Ogdensburg				
	Pontiac	324 Isabella St.	500	1939–55
	Star		200	1926–29
	Strand	Caroline St. at Ford St.	1092	1926–66
Old Forge				
	Brown's		550	1924–35
	New			1926–29
	*Strand		1012	1924–present
Oneida				
	Auroca Hall			1929
	Capitol			
	Carroll			
	Elco		400	1925–29
	Kallet	159 Main St.	1220	1939–69
	Madison	Madison St.	960	1920–53
	Oneida Cinema	State St.		
	Regent		600	1925–31
Oneonta				
	Casino			
	Happy Hour			
	Metropolitan			

(Continued on next page)

Town/Location	Theater Name	Street	Seating	Yrs. of Operation
	Oneonta	41 Chestnut St.		1898–2004 (under redevelopment)
	Palace → Maxey	251 Main St.		1920–66
	Showcase Cinema	11 Elm St.	275	1966–2000
	Strand			1915
Oriskany Falls				
	Cross Opera House			1903
	McLaughlin's		450	1913–32
	Star	Main St.	278	1932–52
Orwell				
	Ideal Rest			1925
Oswego				
	Capitol	1100 E. Second St.	920	1924–51
	Gem aka State	71 E. Bridge St.	250	
	Hippodrome	W. Second St.		1929
	Orpheum	W. Bridge St.	650	1935
	*Oswego	138 W. Second St.	1800	1945–present
	Pierce → Strand	147 W. Second St.	1400	1921 / 1924–51
	Richardson	E. First St.	1081	1931–39
Ovid				
	Franklin Hall		450	1926–38
Owego				
	*Tioga	208 Main St.	793	1926–present
Oxford				
	Citizen's Opera House		400	1926–29
	Smalley's		450	1938–55
Palmyra				
	Opera House		700	1925
	Park		700	1927–29
	Strand	Main St.	674	1926–55
Penn Yan				
	Dundee Theater			
	Elmwood	Elm St.	838	1921–70
	Sampson		650	1926–29
Phelps				
	Opera House → Garlock		400	1926 / 1934
	Phelps	Main St.	400	1939–57

(Continued on next page)

Town/Location	Theater Name	Street	Seating	Yrs. of Operation
Philadelphia				
	Crescent	Main St.	283	
Phoenix				
	Strand		300	1925–57
Poland				
	Jim's		150	1926
	Lindy		400	1929
	Poland		209	1939–55
Port Leyden				
	Liberty			1925–29
	Port		240	1939–57
Pulaski				
	Hohman Opera House		450	1925–35
	Kallet		680	1939–69
	Temple		565	1925–35
Raquette Lake				
	Boys' Club		200	1929
	Casino		130	1924–29
Red Creek				
	Powers		350	1925–29
Remsen				
	Remsen		400	1929
Richfield Springs				
	Capitol	Main St.	425	1935–68
	Opera House			1926–29
	Smalley's aka Shaul's		350	1929–34
Rome				
	Capitol	218 W. Dominick St.	2096	1928–77
	→*(perf. arts cntr.)			1985–present
	Carroll	114 E. Dominick St.	800	1911
	→ Strand		1000	1921–65
	Casino	100 N. James St.		
	Idle Hour	300 W. Dominick St.		
	Lyric	248 W. Dominick St.		1907
	→ Regent			1920
	→ Family			1921–54
	Sinks Opera House	E.Dominick St.		c. 1880s–1903
	Star	128 N. James St.	500	1913–28
	Strand	226 W. Dominick St.	1290	

(Continued on next page)

Town/Location	Theater Name	Street	Seating	Yrs. of Operation
	Washington St. Opera House			1893–1902
Romulus				
	Grange		200	1926–29
Sackets Harbor				
	I.O.O.F. Hall		400	1929
	Madison Barracks		460	1926–39
Salisbury Center				
	Pine Crest		400	1925–29
Sandy Creek				
	Allen			1926–29
Saranac Lake				
	Pontiac	Broadway	1200	1924–66
	New		800	1927–29
	Queen		400	1929
	Saranac Inn		300	1924–35
Savannah				
	Opera House		300	1925
Schoon Lake				
	Community		250	1929
	Paramount		250	1939–51
	Strand		500	1924–55
Seneca Falls				
	Fisher		987	1925
	→ Strand	130 Fall St.		1928–55
	Seneca		500	1925–29
Sharon Springs				
	Smalley's	Main St.	350	
Sherburne				
	Sherburne	Main St.	500	1925–53
	Smalley's		600	1929
Sherrill				
	Community		400	1924–29
Skaneateles				
	Huxford	64 Genesee St.	330	1925
	→ Colonial			1942–76
Skaneateles Falls				
	St. Brigid Hall			
Smyrna				
	Smyrna		400	1925–27

(Continued on next page)

Town/Location	Theater Name	Street	Seating	Yrs. of Operation
Sodus				
	Opera House		810	1925–27
	Sodus	Main St.	650	1945–55
Sodus Point				
	Arcade		700	1935–39
	Crescent		90	1925–29
South New Berlin				
	Baptist Church		800	1926–29
South Otselic				
	Dew Drop		100	1925–29
Solvay				
	Allen	408 First St.	500	1914–60
	Community	1725 Milton Ave.	600	1921–53
	Grand	213 Milton Ave.		1912–16
	New	1301 Milton Ave.		
	Royal	113 Milton Ave.		1914
	Solvay	104 Cogswell St.		1910–14
Stamford				
	Smalley's		795	
Trumansburg				
	Burg	Hector St.	340	1945–56
	Cayuga	Hector St.	390	1942–55
	Park		300	1925–29
	Star		250	1927–29
Tully				
	Auditorium		250	1928–31
Tupper Lake				
	Palace		600	1914–35
	State	106 Park St.	500	1938–57
Tupper Lake Junction (Faust)				
	Lyric		265	1934–45
Union Springs				
	Springport	Cayuga St.		
	Union			1928–31
Utica				
	Aerodrome	81 John St.		1913–14
	Alhambra	108 Bleecker St.	900	1910–31
	Avon	212 Lafayette St.	1562	1916–64
	Carlton	628 Bleecker St.	485	1928–39

(Continued on next page)

Town/Location	Theater Name	Street	Seating	Yrs. of Operation
	Columbia	930 Whitesboro St.		1916
	Cornhill1	505 Neilson St.	500	1917
	→ Liberty			1926
	Criterion			1922
	DeLux	Park Ave.	1200	1916–35
	→ Oneida Square			1935–55
	Dreamland	324 Varick St.		1912–16
	Family	896 Bleecker St.	1000	1915–53
	Gem	215 Bleecker St.	213	1911
	→ Empire			1913
	Happy Hour	926 Columbia St.		1917
	Hibernian	869 Bleecker St.		1914
	Highland	1708 Whitesboro St.	678	1925–57
	Hippodrome	31 Lafayette St.	800	
	James	309 James St.	1200	1927–53
	Kallet	1143 Lincoln St.		1928
	Lincoln	617 Cottage Pl.	800	1925–71
	Lumberg	428 Washington		1911
	→ Gaety			1921
	→ Fox			1930
	→ Utica		2000	1931–53
	Lyric	725 Varick St.	1000	1910–35
	aka Rivoli, Sunset,			1935–closing
	Robbins			
	Majestic	110 Lafayette St.	1500	1920–69
	Olympic	127 Lafayette St.	1500	1927–69
	Park	327 Bleecker St.	900	1921
	→ State			1925–31
	Rialto aka Forum	812 Nichols St.	870	1921–55
	Savoy	634 Bleecker St.		1914
	Schuyler	1131 Whitesboro St.		1914
	→ Plaza			1915
	Sheridan	Yorkville St.		1915
	Shubert	203 Bleecker St.	1200	1910
	→ Buckley			1914
	→ Colonial			1917–45
	South	264 South St.	1100	1914
	→ Orpheum			c. 1930–53
	Stanley	259 Genesee St.	2945	1928–1974
	→*(perf. arts cntr.)			1974–present
	Star	644 Bleecker St.		1911
	State	Bleecker St. at John St.		?–1926

(Continued on next page)

Town/Location	Theater Name	Street	Seating	Yrs. of Operation
	Strand (Stanley)	259 Genesee St.	2963	1927–69
	Sunset	725 Varick	575	1942–45
	Theatorium	206 Genesee St.		1909
	*Uptown	2014 Genesee St.	1038	1927–present
	Utica	428 Washington St.	2000	1931–53
Vernon				
	Union Hall			1925
	Vernon		210	1939–43
Walton				
	Majestic			
	Smalley's		750	1929–57
	Walton		400	1914
	→*(perf. arts cntr.)			?–present
Warsaw				
	Farman's	Main St.	1000	1931–66
	O-At-Ka	Main St.	444	1926–53
Waterloo				
	Star		500	1925–29
	State		424	1934–55
Watertown				
	Antique			1908
	→ Liberty	239 Court St.	1000	1921–59
	Olympic	224 State St.	2300	1920–66
	Opera House	152 Arsenal St.	1792	
	→ Avon			c. 1920–64
	Palace	56 Public Square	800	1920–55
	Strand	136 Franklin St.	400	1935–55
	Victoria	107 Public Square	600	1924–55
Waterville				
	Lyceum		300	1924–31
	Opera House	Main St.		1880
	→ Strand		400	1935–67
Watkins Glen				
	*Glen	112 N. Franklin St.	632	1924–67
Weedsport				
	Burritt Opera House	50 S. Seneca St.	600	1995–35
	→ Weedsport		500	1936–62
West Winfield				
	Bisby Hall		350	1926–51

(Continued on next page)

Town/Location	Theater Name	Street	Seating	Yrs. of Operation
	People's		500	1929–31
	Point	Main St.	335	1939–55
	Tyler		300	1925–29
Wolcott				
	Bijou		250	1925–29
	*Palace	61 E. Main St.	340	1932–present